AF602047

Microbiology of Fish and Fishery Products

The Authors

F Parthiban (1980) is the Assistant Professor of Tamil Nadu Fisheries University, Nagapattinam. He has 10 years of fisheries industrial experience as managers in aquaculture, fish processing, and quality control at Saudi Arabia, Nigeria, and Tanzania and 4 years of experience in teaching, research and extension programmes. He has Recipient of CIFE, ICAR fellowship for Master of Fisheries Science in Post Harvest Technology at Central Institute of Fisheries Technology. He has successfully completed internal auditor and lead auditor courses for Food Safety Management System, ISO 22000:2005. He is the member of society of Fisheries Technologist and Professional Fisheries Graduates Forum in India. In addition, he has participated in many National and International Conferences, Symposia, Workshops, Seminars, etc., organized by fisheries development organizations. He has good expertise in aquatic food processing, aquatic food safety and quality. He has been handling courses for fisheries graduate on Fundamentals of Biochemistry , Biochemical Techniques and instrumentation, Microbiology of fish and fishery product, Fish Processing Technology. He has been conferred with "Best Teacher Award" by the Tamil Nadu Fisheries University for the year 2017. He has published research and extension articles in national and international journal.

Dr S Felix, Ph.D., Vice Chancellor, Tamil Nadu Dr. J. Jayalalithaa Fisheries University, Nagapattinam is having experience of more than 35 years in the area of fisheries and aquaculture in teaching, research, extension and administration. He has organized more than 20 Nos. of seminar/conference/symposium at National and International levels. Currently he is also the President (2018-19) of World Aquaculture Society, Asian Pacific Chapter and Chairman, ICAR's BSMA Committee (2018-19). He has more than 60 research papers published in National and International journals. He has authored more than 10 books and 25 manuals. He is specialized in the area of advanced systems in aquaculture and aquariculture such as raceway, biofloc technology, RAS, etc,. He has operated around 20 externally funded research projects.

Microbiology of Fish and Fishery Products

by

F Parthiban

S Felix

Tamil Nadu Dr. M.G.R. Fisheries College & Research Institute
Tamil Nadu Dr. J. Jayalalithaa Fisheries University
Ponneri, Tiruvallur District, Tamil Nadu

DAYA PUBLISHING HOUSE®
A Division of
ASTRAL INTERNATIONAL PVT. LTD.
New Delhi – 110 002

ISBN: 9789388173261 (Int. Edition)

Published by : **Daya Publishing House®**
A Division of
Astral International Pvt. Ltd.
– ISO 9001:2015 Certified Company –
4736/23, Ansari Road, Darya Ganj
New Delhi-110 002
Ph. 011-43549197, 23278134
E-mail: info@astralint.com
Website: www.astralint.com

Laser Typesetting : **Classic Computer Services,** Delhi - 110 035

Printed at : **Neelam Graphics, Delhi - 110007**

डॉ. जे. के. जेना
उप महानिदेशक (मत्स्य विज्ञान)

Dr. J. K. Jena
Deputy Director General (Fisheries Science)

भारतीय कृषि अनुसंधान परिषद
कृषि अनुसंधान भवन-II, पूसा, नई दिल्ली 110 012
INDIAN COUNCIL OF AGRICULTURAL RESEARCH
KRISHI ANUSANDHAN BHAVAN-II, PUSA, NEW DELHI - 110 012

Ph. : 91-11-25846738 (O), Fax : 91-11-25841955
E-mail: ddgfs.icar@gov.in

Foreword

Microorganisms in their natural habitat play an important role in cycling of nutrients in the ecosystem. In the process of performing their primary role in nature, the microflora associated with food cause spoilage of foods. Thus, the knowledge on types of microorganisms naturally associated with plant and animal foods helps to predict the microbial types that could be present at later stages of handling, storage and preservation of food. It is a well-known fact that seafood is an important contributor to global food and nutritional security. Fish, therefore, is considered an irreplaceable animal-source, providing essential nutrients of high bioavailability, including quality animal protein, essential fats, minerals and vitamins. However, poor handling of these fish in the landing centres, long-distance transportation of fish without ice, adoption of unhygienic curing methods, poor packaging and improper storage often have led to high degree of contamination of the fish with harmful microorganisms and also spoilage, leading to enormous health risk to the consumers. Further, such spoilage also results in heavy economic loss to the fisherman and also the traders.

I am happy to note that the authors of the book '***Microbiology of Fish and Fishery Products***' have tried to include all relevant aspects of microbiology associated with the fish production and post-harvest value-chains, including the general introduction about the historical evaluation of basic microbiology and food microbiology; aspects of intrinsic and extrinsic parameter associated with the microbial growth; spoilage and association of pathogenic bacteria in chilled, frozen, cured, thermal processed and value added products; fungal, parasitic and viral pathogens associated with fish and fishery products; and conventional and advanced techniques followed in microbial analysis.

I congratulate the authors for their sincere efforts in bringing out this publication.

(J.K. Jena)

Message

This book serves as a general practical guide for graduate students in food microbiology and hand book for technologist and supervisor working in the food processing industries. It is focusing on basic skill building in enumeration and isolation of bacteria of public health significance associate with fish and fishery products. It is the need of the hour to improve the overall food safety and quality system through an adequate and appropriate understanding of the quality and safety parameters by all the stakeholders in the food chain. To achieve this objective, this book has been designed. It describes the methods for detection of food–borne pathogens in the food processing laboratory and students in the field of microbiology to culture and biochemical methods to confirm. This book gives details on the microbiology of chilled fish, frozen fish thermal processed and value added fishery products individually.

I hope this book will be useful for students, food technologist and researchers in the field of microbiology of fish and fishery products.

S. Felix
Vice Chancellor
Tamil Nadu Dr. J. Jayalalithaa Fisheries University

Preface

Indian fish production through wild catch and aquaculture has steadily increased to approximately 11.41 million metric tonnes in the years 2016-17. During the year 2016-17, a sum of Rs. 37,870.90 crore has been earned as foreign exchange through the export of Rs.11.34 lakh tonnes of fish products. Tamil Nadu has exported Rs. 5,308 crore worth seafood products in this year. Food safety is a concern facing the seafood industry today. There are extensive coverage in daily press on the food safety issues such as *Salmonella, Vibrio cholerae, Listeria monocytogenes*, transfer of antibiotic resistant microorganisms, quarantine issue related to viral pathogens, *etc.* that threatening the consumers. It is very essential to produce Fisheries Graduates with adequate knowledge on the above said aspects to tackle such food safety issues.

Microbiology of Fish and Fishery Products is a course in the curriculum of B.F.Sc. This manual provides detailed information on factors affecting microbial growth, spoilage and public health significance bacteria in fish and fishery products. It is also providing information on other microbial hazards such as viruses and shellfish poisoning. The techniques for isolation, identification and characterization of various microorganism in fish and fishery products dealt with this manual. The contents of this manual will serve as an excellent study reference material for B.F.Sc., M.F.Sc., and Ph.D students for field and research purpose.

Authors

Contents

Chapter 1

Introduction to Food Microbiology

1.1. Introduction

Seafood is an important contributor to global food and nutritional security. FAO estimates indicate that fish provides more than 2.9 billion people with almost 20 per cent of their intake of animal protein. A portion of 150 g of fish can provide about 50–60 per cent of an adult's daily protein requirements. Fish can therefore be considered an irreplaceable animal-source, providing essential nutrients of high bioavailability which are found in limiting amounts in the diet. These nutrients include animal protein, essential fats, minerals and vitamins. Fish and fishery products have a fairly good safety record. About 40 per cent of fishborne illnesses were due to scombrotoxic fish poisoning, 20 per cent due to ciguatoxic fish poisoning, and the rest due to microbiological causes. Aquatic foods contain a variety of microorganisms from different sources. These contaminants in food cause problems of spoilage and health risk to consumers.

Fish are not sterile, but inner tissue of healthy fish is sterile and the microflora includes natural flora of the waters from which fish are harvested and acquired transient flora from environment especially during handling, processing, storage *etc.* Transient flora include microorganisms entering the food from fish contact surface (crafts/gears/baskets/fish holds *etc.*), the air, soil, water/ice used for washing, food handlers, packaging material and storage environment. Microorganisms are mainly associated with outer slime, gills and intestine. Microbial load is higher in intestine followed by gills and skin. Natural microflora of fish varies depending on the habitat of fish (freshwater/marine/river/water/lake/*etc.*), its feeding habit and life history stages. Generally, fresh warm water fish have more mesophilic bacteria than cold water fish. Fish harvested from polluted water contain variety of microorganisms depending on nature of pollutant, and also human pathogens such as bacteria, fungi, viruses, protozoans, parasites *etc.*

Once the fish dies the associated microorganisms affect quality due to spoilage. Microorganisms like bacteria are present in live fish on the skin, gills and in gut, but these are commensal bacteria and most of these are not human pathogens. But some naturally occurring marine bacteria like *Vibrio parahaemolyticus* and *V. vulnificus* may cause illness in man. Some naturally occurring bacteria such as *Clostridium botulinum* may be found in fish gut and if anaerobic conditions are provided, they could multiply, produce toxin that can be even lethal. Thus, it is necessary to maintain the quality by destroying associated microorganisms and preventing growth of surviving microorganisms. The different preservation methods mainly aim at maintaining fish quality by reducing, killing or inactivating associated spoilage microorganisms.

1.2. History of Microorganisms in Foods

The role of microorganisms in spoilage and food poisoning was realised only after the establishment of bacteriology or microbiology as a science during the early 18th century. However, adoption of several approaches to avoid/reduce spoilage was known since early civilization. The use of salted meat, fish, fat, dried animal skin and cereals was practiced by the Sumarians at about 3000 B.C. Romans were known for adopting preservation techniques for meat and also use of snow for perishables including prawns at around 1000 B.C. Smoking of meats and making of cheese and wines emerged during the same time. But the people at that time were not fully aware of nature of the preservation techniques and also the role of foods in the transmission of diseases and the danger of eating infected animal meat.

A Kircher, a monk, was first person in 1658 to observe the decaying bodies, meat, milk and other substances and believed it to be due to invisible worms, thus suggesting the role of microorganisms in food spoilage. During that time the theory of spontaneous generation of life was gaining importance and L. Spallanzani in 1765 was first to oppose it with his experiment using beef broth which remained sterile after boiling and sealing. But he was not able to convince the proponents of spontaneous generation who felt that air was necessary for life to begin. Later in 1837, Schwann's experiment of passing heated air in to the boiling broth caused it to remain sterile thus disproving spontaneous generation of life. But it was not until Pasteur's well planned swan neck flask experiment that the theory of spontaneous generation of life was put to rest.

The development of canning of meat in glass bottles by Nicholas Appert in 1809 had practical application in food preservation. Pasteur was the first person to appreciate and understand the presence and role of microorganisms in food. He demonstrated the role of microorganisms in souring of milk in 1837, and applied heat to destroy undesirable microorganisms in wine and beer, in about 1860.

Involvement of microorganisms in food spoilage and food poisoning has led to the development of preservation methods to arrest and kill microorganisms. The science of food microbiology has application in all kinds of food meant for human consumption and aims at ensuring the wellbeing of consumers.

1.3. Role and Significance of Microorganisms in Foods

1.3.1. Role of Microorganisms in Nature and in Foods

Microorganisms in their natural habitat play an important role in cycling of nutrients in the ecosystem. In the process of performing their primary role in nature, the microflora associated with food cause spoilage of foods meant for human consumption. Thus, the knowledge on types of microorganisms naturally associated with plant and animal foods helps to predict the microbial types that could be present at later stages of handling, storage and preservation of food.

Depending on the nature of food and its habitat a variety of microorganisms are expected in the food and these may often affect the safety of food. Therefore, information on factors such as total number and types of microorganisms naturally present, types of microorganisms present in specific food, and ones which are not natural to the food becomes necessary. This information becomes valuable in ascertaining the safety of food during different stages of processing, handling and storage.

All surface waters such as ponds, lakes, rivers and oceans differ in their physical, chemical and biological characters. Depending on the nutrient status of water body, the microbial load varies with higher numbers encountering in eutrophic waters. Ground waters or subterranean waters generally have very low microbial load because of filtration effect of soil layers.

Categories of Microorganisms in Natural Waters

The natural waters contain a variety of microorganisms. These include,

- Natural flora: Microorganisms natural to the water body and
- Transient flora: Microorganisms entering the water body from outside environment like from soil, air and through pollutants.

Microorganisms in natural aquatic environment play an important role in nutrient recycling, and as primary producers and decomposers of organic matter. All the microorganisms present in a water body can be seen as surface flora of inhabiting organisms. These not only include spoilage organisms but also human pathogenic microorganisms especially in sewage contaminated waters.

Primary Source of Microorganisms found in Food

The foods of plants and animal origin carry several microorganisms associated with their natural habitat. Plants carry typical micro-flora on their surface and also get contaminated from outside sources. Animals carry microorganisms on their surface and intestine, and also contain contaminants from surrounding environment. Through their excretions and secretions animals release microorganisms in to surrounding environment. Besides, both plants and animals carry pathogenic microorganisms capable of causing human illness. The food associated microorganisms are influenced by the availability of specific nutritional requirements and the environmental parameters. The primary sources of entry of microorganisms in to foods are from the soil, water, air, during handling, processing transportation and storage of foods.

Soil

Soil being the rich source of several kinds of microorganisms immediately contaminates the plants and edible plant parts, and the surface of animals with the soil associated microorganisms. As the soil particles are carried in to aquatic environment through wind, rain and other means contamination of water takes place with several soil micro-flora. Therefore, it is not uncommon to find several microorganisms both in soil and water environment. These soil derived microorganisms form part of the microbial flora involved in spoilage of foods of plant and animal source. Thus, there is a need to reduce the load of soil microorganisms in foods which can be achieved by removing the soil by washing the surface of foods with good quality water, and by avoiding contact with soil/dust.

Water

Natural waters not only contain several microorganisms native to the aquatic environment but also from soil, raw/treated sewage and pollutants entering the water body. The microbial numbers and types vary in different water bodies depending on the nutrient status. Thus, all kinds of microorganisms found in water are likely to be associated with the aquatic organisms as surface flora. Use of such water for food processing will add microorganisms from water to food. Sewage waters containing human pathogenic microorganisms contaminate foods when such waters are used without proper treatment. The water used in food processing should meet agreeable chemical and bacteriological characteristics.

Air

Air contains several microorganisms which may get deposited on the food being processed and handled. Though the air does not contain natural flora of microorganisms, whatever microorganisms encountered are those associated with the suspended solid material and water droplets. The sources of microorganisms to air are from dust, dry soil, and water spray from natural surface waters, droplets of moisture from coughing, sneezing and talking by food handlers, from sporulating moulds growing on walls, ceilings, floor, foods and food ingredients. Thus, it is likely that the microorganisms persisting in air get deposited on the food being processed and contribute for microbial load and subsequent spoilage of food. The number of microorganisms present in air depends on factors such as extent of movement of air, sunshine, humidity, location and amount of suspended dust in air. Quiet air allows settling of microorganisms but the moving air brings in microorganisms and keeps them suspended. Thus, the number of microorganisms in air is increased by air currents caused by movement of people, by ventilation and by breeze. The rain or snow removes microorganisms from the air.

Micro-flora of Food Processing Facility

The nature of micro-flora in a food processing facility varies depending on the nature of food being processed. Hence characteristic microbial populations are encountered in different processing units. Also variations may be observed in microbial numbers from one area of processing plant to another. The microbial types present inside the processing plant are related to quality of air outside the plant and the microbial population levels are related to the level of activity of workers.

Reducing Microbial Load in Processing Area

There is a need to reduce microbial load in the processing area. This can be achieved by installing filtration, chemical treatment and heat or electrostatic precipitation units, and taking measures in preventing the build up after reducing the microorganisms. The build up of microorganisms in the processing area can be prevented by maintaining the positive pressure in food process area, installing filters in ventilating systems that prevent spread of microorganisms from one part of a plant to another and installing UV- irradiated air locks at doors to reduce the number of organisms carried by workers.

Handling and Processing

Foods grown/cultured in natural environment containing specific groups of microorganisms are further contaminated by several microorganisms during harvesting, handling and processing. Further, addition of microorganisms to food may take place from;

- All food contact surfaces including equipments coming in contact with foods, packaging material, and from food handlers. Foods are also prone for microbial contamination during transportation and storage.
- Use of sewage contaminated water for washing foods being processed contaminates it with human pathogenic microorganisms
- All the microorganisms associated with food handlers enter the food during handling of food from hands, garments, body surface, hair *etc.* under poor personnel hygiene practices.

Therefore, it becomes necessary to prevent/reduce contamination and microbial build-up in foods during handling and processing so as to produce safe food with good keeping quality.

1.3.2. Significance of Microorganisms in Foods

Microorganisms associated with food derive energy from food for those cell growth, maintenance and reproduction. Based on their function microorganisms associated with foods may be divided in to three general groups;

- ☆ Those causing spoilage or undesirable changes in the food
- ☆ Those producing desirable changes
- ☆ Those producing disease

Based on the extent of stability to microbial invasion foods may be classified as;

- ☆ Perishable foods Ex. Fish and meat
- ☆ Semi-perishable foods Ex. Potatoes and Tomato
- ☆ Stable foods Ex. Cereals, Flour and Sugar

The stable or semi-stable foods become unstable or perishable when the moisture content increases.

Factors Affecting Microorganisms in Foods

The survival and activity of microorganisms in foods depends on several factors namely, numbers and types of microorganisms present, type of food, treatments to which the food has been exposed, processing or storage treatments that the food receives, whether the food is to be consumed as it is or heated.

The food associated microorganisms may have useful function, cause spoilage, cause health hazard and play no role or remain inert. The inert microorganisms are those which do not find food environment favorable for their growth, and remain dormant without causing any changes in food.

Causes for Spoilage of Food

Spoilage of food usually occurs due to,

- ☆ Undesirable changes brought about by the microorganisms in the odour, colour, taste, texture and appearance of the food.
- ☆ Some microorganisms may not directly involve in spoilage but bring about changes in food that will facilitate growth of spoilage organisms. Ex. Bacteriophage attacking useful organisms and facilitating growth of undesirable organisms leading to spoilage.

Useful organisms are those which by their activity or fermentation reactions facilitated by the microbial enzymes produce desirable changes in food (Ex. bakery, dairy, wine industries).

Significance of Microbes in Aquatic Food

Microorganisms are the main cause of spoilage in fresh and lightly preserved aquafoods. Additionally, food-borne pathogenic bacteria can occur on aquafoods causing food-borne infection and intoxications. Microorganisms associated with aquafoods can be accommodated into: Indigenous microbiota, Contamination (exogenous) microbiota and Spoilage microbiota.

a. Indigenous Microbiota

The indigenous microbiota of aquafoods consists mainly of bacteria that can cause spoilage while some can also be human pathogens. Natural bacterial populations can be found in skin, gills and the digestive tract. Gram negative bacteria, usually psychrotrophs, predominate especially in fish from cold or more warm temperate waters, while higher amounts of Gram positive bacteria can be found in fish from tropical waters. Pseudomonas, Pyschrobacter, Moraxella, Acinetobacter, Flavobacterium, Vibrio, Photobacterium and Aeromonas are the most common Gram-negative genera that make up initial microbiota while the most common Gram-positive genera are *Micrococcus*, Corynebacterium, Bacillus and Clostridium. Over the last few decades naturally-occurring bacteria have become the leading cause of shellfish-borne illness of known aetiology. Most of these indigenous bacteria belong to the family Vibrionaceae which includes the genera Vibrio, Aeromonas and Plesiomonas.

b. Contamination (Exogenous) Microbiota

Contamination with non-indigenous microorganisms may occur during processing. Microorganisms from the environment, food contact surfaces and workers can be transferred to food. It is known that humans carry many bacteria of which some may be food-borne pathogens such as *Staphylococcus aureus*. Enterobacteriaceae are commonly found in any environment, while contamination with enteric pathogens can also occur in poor hygienic conditions. *Listeria monocytogenes* can attach on food contact surfaces forming biofilms and contaminating seafood.

c. Spoilage Microbiota

In fresh or lightly preserved seafood, the main spoilage mechanism is the growth and metabolic activity of microorganisms, which results in the production of various metabolites that directly affect the organoleptic properties of the product. The main microorganisms responsible for the production of off-flavours/odours (chemical spoilage indices) which lead to seafood rejection are specific spoilage organisms (SSOs). Accumulation of microbial metabolites at levels high enough to cause organoleptic rejection usually occurs when the total microbial population of population of SSOs reach the level of 107-109cfu/g. The time required for microorganisms to reach the minimum spoilage level determines shelf life. Shelf life usually correlates well with the population of total or spoilage microorganisms. *Pseudomonas* and *Shewanella* species are the spoilers of marine fish and crustaceans stored aerobically at low temperatures.

d. Pathogenic Microorganisms

Several bacterial pathogens have been implicated in seafood-borne infections or intoxications. These bacterial pathogens can originate from indigenous microbiota or contaminate fish when they are still in water and/or during processing. There are many pathogenic bacteria that are associated with faecal contamination of seafood include *Salmonella, Shigella, Campylobacter, Yersinia, Listeria, Clostridium, Staphylococcus* and *Escherichia coli.*

Enteric viruses such as hepatitis A virus (HAV), Norwalk-like viruses and Astrovirus are important human pathogens and they are sometimes associated with seafood, especially with shellfish. Shellfish such as oysters, cockles and mussels are filter feeding organisms. Viruses such as the small round structured viruses (SRSV) are an emerging concern associated with the consumption of shellfish. Better protection of harvesting waters is imperative as many outbreaks have been associated with cooked or depurated products.

e. Beneficial Microorganisms

Among the beneficial microorganisms, the Lactic Acid Bacteria (LAB) is most important. The role of LAB in marine products is complex, depending on the fish species, treatment and storage conditions, bacterial species and strains and interaction between bacteria. Sometimes, LAB have no particular negative effect, but in certain cases they are responsible for strong sensory degradation, leading to rejection of the products. The use of LAB in the fish industry is not extensively developed, except in Asia for preparation of fish sauces and traditional food with fermented mixture of fish and vegetable. In most cases, fermentation is due to LAB naturally present in fish or in the carbohydrate added source (vegetables, garlic, *etc.*) and no selected starter is added to control the fermentation. Recently bioprotective potential of endogenous LAB in relation to pathogens is highlighted. Efforts are being made to exploit this ability to control the quality and safety of marine products. However, this technology is still in its infancy compared to dairy products. Some *Lactobacillus* and *Carnobacterium* strains are easy to cultivate and resistant to various conditions. Their production on fish protein hydrolysate and ingestion as dietary supplement may combine the beneficial effect of fish and LAB.

Chapter 2

Parameters Affecting Microbial Growth

Foods meant for human consumption are rarely sterile, and contain several microorganisms. Microorganisms associated with food include natural micro-flora of raw material and organisms that are introduced during harvesting/handling, processing, storage and distribution.

The microbial load in food depends on factors such as

- ✰ Nature of the food
- ✰ Storage environment
- ✰ Properties of organisms in foods
- ✰ Effects of processing

Generally, micro-flora associated with food has no effect and food is consumed without objection and no adverse consequences. But in some instances associated microorganisms manifest their presence leading to spoilage, food borne illness or Transformation of properties of food in a beneficial way (Ex. Fermentation). Foods require time/temperature control during storage, distribution, retail sale and food service.

Factors affecting Microbial Growth in Foods

Microorganisms associated with food and their growth is affected by several factors. These are broadly grouped into 2 types, such as Intrinsic factors and Extrinsic factors. Intrinsic factors are those that are characteristic of the food itself; and extrinsic factors are those that are characteristic to the environment surrounding the food.

2.1. Intrinsic Factors

Intrinsic factors: Includes physicochemical properties of food that affect the microbial growth such as:

- Water activity/moisture content
- pH and buffering capacity
- Nutrients
- Biological structure
- Redox potential
- Antimicrobial constituents
- Competitive microorganisms

i. Water Activity

Moisture content of the food affects microorganisms in foods, and the microbial types present in foods depends on the amount of water available. Water requirement for microorganisms is described in terms of water activity. The water activity (a_w) of a food describes the "bound water" that is available to participate in chemical/biochemical reactions, and to facilitate growth of microorganisms. a_w in the environment and is defined as the ratio of the water vapor pressure of food substrate to the vapor pressure of pure water at same temperature.

$$a_w = P/P_o$$

P = Vapour pressure of water in substrate

P_o = Vapour pressure of solvent (pure water)

a_w is related to relative humidity (RH)

$$RH = 100 \times a_w$$

Water Activity of Solutes and Requirements of Certain Microorganisms

- a_w of pure water: 1.0
- NaCl solution (22 per cent): 0.86
- Saturated NaCl solution: 0.75
- Bacteria generally require higher value of a_w than fungi
- G – ve bacteria require higher a_w than G +ve bacteria
- Most spoilage bacteria do not grow at a_w below 0.91
- Spoilage molds grow at a_w of 0.80
- Halophilic bacteria grow at a_w of 0.75
- Xerophilic and osmophilic yeasts grow at of 0.61

Microorganisms like halophiles, osmophiles and xerophiles grow better at reduced a_w. Microorganisms can not grow below a_w 0.60, and in such situations spoilage of food is not microbiological but due to chemical reactions (Ex: oxidation). Gram negative bacteria are generally more sensitive to low a_w than gram positive bacteria.

Relationship between a_w, Temperature and Nutrition

- Growth of microorganisms decreases with lowering of a_w
- The range of a_w at which the growth is greatest occurs at optimum temperature for growth
- The presence of nutrients increases the range of a_w over which the organisms can survive

pH and Buffering Capacity

The acidity and alkalinity of an environment affects growth and metabolism of microorganisms as the activity and stability of macromolecules, enzymes and nutrient transport is influenced by pH. Generally, bacteria grow fast at pH 6-8. But bacteria that produce acids have optimum pH between pH 5 and 6 (Ex: Lactobacillus and Acetic acid bacteria). Yeast grows best at pH 4.5-6.0 and Fungi at 3.5 - 4.0. In low pH foods (Ex. Fruits), spoilage is mainly by yeasts and fungi than bacteria. Fishes with pH around neutrality (6.5-7.5) favour bacterial growth and spoil rapidly than meat (pH: 5.5 - 6.5). Ability of low pH to restrict microbial growth has been employed as a method of food preservation (Ex: use of acetic and lactic acid). In general, pathogens do not grow at pH levels below 4.6. A pH of 4.6 is appropriate to control spore-forming pathogens. A pH minimum of 4.2 is appropriate to control *Salmonella* spp. and other vegetative pathogens.

Buffering capacity refers to the ability of foods to withstand pH changes. Microorganisms have ability to change pH of the surrounding environment to their optimal level by their metabolic activity. Decorboxylation of aminoacids releases amines which increases surrounding pH. Deamination of aminoacids by enzyme deaminases release organic acids causing decease in pH. Thus, protein rich foods like fish and meat have better buffering capacity than carbohydrate rich foods.

Fermentation is used to preserve the fish by increasing the acidity of foods using organic acids. The pH interacts with other factors such as $a_{w,}$ salt, temperature, redox potential, and preservatives to inhibit the growth of pathogens. It also has an impact on the lethality of heat treatment of the food. For instance, less heat is needed to inactivate microbes at low pH. The acidity is also used as a control mechanism in food due to its buffering capacity. Foods with a low buffering capacity will change pH quickly in response to compounds produced by microorganisms. But, fish meat has more buffering capacity than vegetables due to the presence of proteins. **Titratable acidity** (TA) is a better indicator of the microbiological stability of foods than pH.

ii. Nutrient Content

Microorganisms require basic nutrients for growth and maintenance of metabolic functions. They are water, an energy source, nitrogen, vitamins, and minerals. Nature of nutrient availability favors growth of different microorganisms. Microorganisms derive energy from carbohydrates, alcohols, and amino acids. They utilize simple carbohydrates and amino acids first and then more complex forms of nutrients. Ability of microorganisms to utilize nutrients favors their growth on foods, while inability to utilize nutrients reduces microbial growth. Ex. Saccharolytic microorganisms grow well on cereals using carbohydrates as carbon source.

- Proteolytic bacteria grow well on fish and meat.
- Diverse microbial group are associated with fruits containing variety of sugars (sucrose, fructose *etc.*) favoring a variety of microorganisms and yeasts.
- Generally, foods where nutrients are easily available (fish/meat) support high microbial load and activity. Foods with low water content are protected against microbial invasion.

Gram positive bacteria are more fastidious and are not able to synthesize certain nutrients required for growth. For instance, *S. aureus* requires amino acids, thiamine, and nicotinic acid for their growth. Gram negative bacteria can derive their nutritional requirements from carbohydrates, proteins, lipids, minerals, and vitamins. Some pathogens require specific nutrients. For instance, *Salmonella enteritidis* needs availability of free iron.

iii. Biological Structure

Biological structures such as skin of fish or shell of shrimps/molluscs may prevent the entry and growth of pathogenic microorganisms. But, some pathogenic microorganisms may be present in the skin, gills and gut of raw aquatic foods. The penetration of microorganisms through these barriers is influenced by several factors. Physical damage due to handling during harvest, transport, or storage allows the penetration. Processing steps like gutting, beheading, peeling, de-skinning, *etc.* can also destroy the barriers. Heat processing can also break down certain protective biological structures.

iv. Redox Potential

Oxidation-reduction or redox potential is defined in terms of the ratio of the total oxidizing (electron accepting) power to the total reducing (electron donating) power of the substrate. Microorganisms show varying degree of sensitivity to O-R potential (Eh) of growth medium. Redox reaction occurs as a result of transfer of electrons between atoms or molecules. O-R potential of a substrate is defined as the ease with which the substrate loses or gains electrons. Substrate is oxidized when it loses electrons and is called as good reducing agent. Substrates that gain electrons become reduced and are a good oxidizing agents.

$$Cu \rightarrow Cu + e$$

Oxidation may also achieved by addition of oxygen.

$$2\,Cu + O_2 \rightarrow 2\,CuO$$

Transfer of electros from one compound to another creates a potential difference between two compounds and is measured by an instrument and expressed as millivolts (mv). Highly oxidized substances have more electric potential (positive potential) and reduced substances negative electrical potential. Zero electric potential when oxidation and reduction are equal. O/R protential of a system is expressed by Eh. Aerobic microorganisms require positive Eh for growth and anaerobes, negative Eh. Reducing conditions in food is maintained by -SH groups in meat and ascorbic acid and reducing sugars in fruits and vegetables.

Factors Influencing O/R Potential of a Food

- ☆ The characteristic of O/R potential of the original food.
- ☆ Poising capacity - (Resistance to change in potential of food)
- ☆ Oxygen tension of the storage atmosphere of food
- ☆ Access that the atmosphere has to the food
- ☆ Microbial activity

Eh Requirement of Microorganisms

- ☆ Aerobic microorganisms require oxidized condition (+ Eh) for growth. Ex. *Bacillus* sp.
- ☆ Anaerobes require reduced condition (-Eh). Ex. *Clostridium* sp 11
- ☆ Microaerophils are aerobes growing at slightly reduced condition. Ex. *Lactobacillus, Campylobacter.*
- ☆ Facultative anaerobes have capacity to grow both under reduced and oxidized condition. *e.g.* Yeasts.
- ☆ Plant foods have positive Eh (fruits, vegetables) and spoilage is mainly caused by aerobes (bacteria and molds).
- ☆ Solid meat and fish have negative Eh (-200 mv), and minced meat positive Eh (+200 mv).

The Eh range required for the growth of microorganisms are for aerobes +500 to +300 mV, facultative anaerobes +300 to -100 mV and anaerobes +100 to less than -250 mV. For instance, *C. botulinum* is a strict anaerobe that requires <+60mV for growth. The relationship of Eh for growth is significantly affected by the presence of salt and other food constituents. For instance, in smoked herring, toxin was produced at 15°C within three days at an Eh of +200 to +250 mV.

Presence or absence of appropriate quantity of oxidizing/reducing agents in the medium influences growth and activity of all microorganisms. The Eh values vary depending on the changes in the pH of food, microbial growth, packaging, the partial pressure of oxygen in the storage environment, ingredients and composition. The measurement of redox potential of food is done rather easily by using electrodes. In case of multi-component foods, the redox potential of the surrounding areas and microenvironments should also be measured in addition to each component. Measurement of redox potential alone cannot be taken to evaluate the potential for pathogen growth in foods.

v. Naturally Occurring and Added Antimicrobials

Foods intrinsically contain naturally-occurring antimicrobial compounds that provide some level of microbiological stability. All foods have one or the other mechanism to prevent or limit potentially damaging effects by microorganisms through protective physical barriers to infection (Ex. skin, shell, and husk) and antimicrobial components. Natural covering of some foods provide excellent protection against entry and subsequent damage by spoilage microorganisms. These include outer covering of fruits, outer shell of egg, skin covering of fish and meats.

Physical damage to outer barrier allows microbial invasion and cause spoilage. Some foods are resistant to attack by microorganisms and remain stable due to the presence of naturally occurring substances which have antimicrobial property. Many plant species possess essential oils which are antimicrobial. Fish contains antimicrobial peptides present in the mucus, gills and other external surfaces. In addition, some food processing also results in the formation of antimicrobial compounds. For instance, smoking of fish results in the deposition of antimicrobial substances (phenol) onto the surface. Phenol is not only an antimicrobial, but also lowers the surface pH. Maillard reaction products can also impart some antimicrobial activity.

Some types of fermentations also result in the production of antimicrobial substances, including bacteriocins, antibiotics, and other related inhibitors. Bacteriocins are proteins or peptides that are produced by certain strains of lactic acid bacteria that inactivate most of the bacteria. For instance, lantibiotic nisin is produced by certain strains of *Lactococcus lactis* and approved for food applications in over 50 countries. It is a polypeptide effective against most gram positive bacteria but is ineffective against gram negative organisms and fungi. It is also used to control spore-forming organisms and shows some interactive effect with heat.

vi. Competitive Microflora

The nature of microorganisms (microbial associations) encountered in foods can affect micro-flora of food which are called implicit factors. These include

- ✰ Properties of organisms present in food
- ✰ Response of these organisms to their environment
- ✰ Interaction with other organisms in food

Effect of Activities of Microorganisms on Food Micro-flora

1. Among several organisms present in food, microorganisms which find condition suitable (in food) for growth dominate over other organisms.

Ex. Molds can grow on dry fish, but slowly, than bacteria. In fresh fish bacteria overgrow molds as conditions are most favourable. Faster growing bacterial growth is inhibited by low a_w or low pH where moulds grow and cause spoilage.

2. Some food borne microorganisms produce metabolite/substances such as antibiotics, bacteriocins, hydrogen peroxide, organic acids *etc.* which are inhibitory/lethal to other microorganisms.
3. Spoilage microorganisms can interact wherein growth of one favours the growth of others. Ex: In low a_w food (grain) growth of few molds increases aw leading to growth of other xerophilic molds.
4. One organism may increase nutrient availability to others by degrading complex food substrates to simple forms.

Some microorganisms may remove inhibitory substances thereby permit the growth of others. Within the microbial flora in a food, there are many important biological attributes that influence the individual growth rates of the microbial strains and the mutual interactions or influences among species in mixed populations.

a. Growth

In a food environment, an organism grows in a characteristic manner and at a characteristic rate. The length of the lag phase, generation time, and total cell yield are determined by genetic factors. Accumulation of metabolic products may limit the growth of a particular species. If the limiting metabolic product can be used as a substrate by other species, these may take over (partly or wholly), creating an **association or succession.** A food at any one point in time has a characteristic flora due to interactions between environmental factors and microorganisms, known as its **association.** As the microbial profile changes continuously, one association succeeds another and this is called **succession.** Many examples of this phenomenon have been observed in the microbial deterioration and spoilage of foods.

b. Competition

In food systems, antagonistic processes usually include competition for nutrients, competition for attachment/adhesion sites, unfavorable alterations of the environment, and a combination of these factors. In muscle foods, *S. aureus* is found in low numbers because the *Pseudomonas, Acinetobacter, Moraxella* association grows at a higher rate, outgrowing the staphylococci. Organisms of high metabolic activity consume the required nutrients, selectively reducing these substances, and inhibiting the growth of other organisms. Depletion of oxygen or accumulation of carbon dioxide favors facultative obligate anaerobes which occur in vacuum-packaged fish meats held under refrigeration. *e.g.* Photobacterium phosphoreum.

c. Effects on Growth Inhibition

Changes in growth stimulation have been reported among several foodborne organisms, including yeasts, micrococci, streptococci, lactobacilli and Enterobacteriaceae. There is several growth stimulating mechanisms:

1. Metabolic products from one organism can be absorbed and utilized by other organisms.
2. Changes in pH may promote the growth of certain microorganisms. In fermented foods, acid production establishes the dominance of lactic acid bacteria. In high acid foods, growth of molds raises the pH, thus stimulating the growth of *C. botulinum*
3. Changes in Eh or a_w in the food can influence symbiosis. At warm temperatures, *C. perfringens* can lower the redox potential in the tissues of freshly processed fish so that even more obligately anaerobic organisms can grow.

2.2. Extrinsic Factors

Extrinsic factors include conditions of storage environment that affect the properties of the food as well as the growth of microorganisms such as,

1. Gaseous atmosphere/packaging
2. Time/Temperature of storage
3. Relative humidity

The intrinsic factors of the food are influenced by the conditions of storage environment, and there by affect the quality of food.

i. Gaseous Atmosphere/Packaging

Exposure of foods to gases in the storage environment (gaseous environment) affects growth and survival of microorganisms in foods. Since exposure of food to oxygen favors growth of aerobic microorganisms, gaseous environment need to be modified to ensure reduced microbial activity and resultant spoilage. This approach is commonly employed in the preservation of fruits and vegetables.

Gases Used to Control Microorganisms in Foods

Carbon dioxide, ozone and nitrogen are most important gases used to control microorganisms in food. Several ready to eat foods are packed in the presence of these gases to reduce microbial activity and extend shelf life of packed foods. Such foods are referred to as Modified Atmosphere Package (MAP) foods.

Carbon dioxide is single most important atmospheric gas used to control microorganisms in foods and is used in varying concentration depending on the type of food. Carbon dioxide in elevated pressure is also used in carbonated water and soft drinks. Molds and Gram negative microorganisms are more sensitive to CO_2 than Gram positive bacteria. Lactobacilli are resistant to CO_2. Yeasts show considerable resistance and tolerate high CO_2 level but can cause spoilage of carbonated beverage (Ex. *Brettanomyces* sp.)

Mechanism of Inhibition

Carbon dioxide mainly acts as bacteriostatic agent. But some microorganisms are killed by prolonged exposure. Mechanism of inhibition of CO_2 is due to the formation of carbonic acid which lowers pH. Lowered pH affects physical properties of plasma membrane of microorganisms and affects solute transport, inhibits key enzymes, and reacts with amino group of proteins causing changes in their property and activity. Ozone (O_3) is also has antimicrobial properties and extends shelf life of certain fruits and vegetables foods. O_3 concentration of 0.15-5 ppm is known to double the shelf life by inhibiting spoilage bacteria and yeast.

Gases inhibit microorganisms by two mechanisms. First, they have a direct toxic effect on the growth and proliferation. Carbon dioxide (CO_2), ozone (O_3), and oxygen (O_2) are gases that are directly toxic to certain microorganisms. The inhibition mechanism is dependent upon the chemical and physical properties of the gas and its interaction with the aqueous and lipid phases of the food. Oxidizing radicals generated by O_3 and O_2 are highly toxic to anaerobic bacteria and also have an inhibitory effect on aerobes. CO_2 is effective against obligate aerobes. Second inhibitory mechanism is by the modification of the gas composition, which has indirect inhibitory effects by altering the ecology of the microbial environment. Atmospheres that have a negative effect on the growth of one particular microorganism may promote the growth of another. This effect may have positive or negative consequences depending upon the native pathogenic microflora and their substrate.

A variety of common technologies are used to inhibit the growth of microorganisms. They are modified atmosphere packing (MAP), controlled atmosphere packaging (CAP), controlled atmosphere storage (CAS), direct addition of carbon dioxide (DAC), and hypobaric storage. CAP and MAP use CO_2, N_2 and ethanol for the extension of the shelf life of food. CO_2 dissolves in the food and lowers the pH of the food. An increase in salt concentrations, decrease the CO_2 solubility. N_2, being an inert gas, has no direct antimicrobial properties. It displaces oxygen in the food package either alone or in combination with CO_2, thus having an indirect inhibitory effect on aerobic microorganisms. The combination of gases, CO_2, N_2, and O_2 used for white fish is 40 per cent, 30 per cent and 30 per cent ; while for fatty fish is 60 per cent, 40 per cent and 0 per cent, respectively.

ii. Effect of Time/Temperature Conditions on Microbial Growth

Storage temperature refers to temperature at which the food is handled and stored. Microorganisms growing over a wide range of temperature have been reported from food. The minimum temperature supporting the growth of microorganism was found to be well below the freezing temperature of water, and the highest temperature close to boiling temperature of water. However, no single microorganism is capable of growing over this wide temperature range. Thus, selecting a proper temperature for the storage of different food types helps in maintaing quality. Each microorganism exhibits minimum, optimum and maximum temperature at which growth occurs, called cardinal temperature. This temperature is characteristic for an organism and is influenced by factors such as nutrient availability, pH of growth medium, water activity *etc.*

Grouping of Microorganisms Based on Temperature Requirement for Growth

Table 2.1: Microorganisms are Grouped in to 4 Broad types based on Cardinal Temperature.

Groups	*Mini. Temp. (°C)*	*Opt. Temp (°C)*	*Max. Temp (°C)*
Psychrotrophs	-5 to +5	25-30	30-35
Psychrophiles	-5 to +5	12-15	15-20
Mesophiles	5-15	30-40	40-47
Themophiles	40-45	55-75	60-90

Psychrophiles

Microrganisms capable of growing at low temperature are called psychrophiles. These are further divided in to 2 types based on their optimum temperature for growth. Obligate psychrophiles (cold loving) – These have temperature optima of 12-15°C, but unable to grow above 20°C. These are confined to Polar Regions and deep marine environment.

Psychrotrophs (Facultative Psychrophiles)

These have same minimum temperature for growth as psychrophiles but have higher optimum and maximum growth temperature. Thus are found in most diverse habitats, grow well in refrigerated temperature and cause spoilage of chilled food. Ex: *Alcaligenes, Cornybacterium, Flavobacterium, Lactobacillus, Pseudomonas, Enterococcus etc.* These bacteria grow well at 5-7°C (refrigerator temperature) and cause spoilage of meat, fish, poultry, eggs *etc.*

Mesophiles

These grow well under normal temperature conditions of 30-40°C. As a rule mesophiles grow quickly at their optimum temperature than psychrotrophs. Thus the spoilage of food at mesophilic temperature range is rapid than at chill temperature.

Mesophiles are found in most genera of bacteria.

Thermophiles

These are high temperature loving microorganisms with the optimum temperature in the range of 55-75°C. Thermophiles are generally less important in food microbiology. But thermophilic spore formers of the genus Bacillus and Clostridium cause spoilage of canned foods.

Molds and Yeasts

Like bacteria, many molds and yeasts are associated with food. Many spoilage molds are able to grow over a range of temperature. Some molds grow at refrigerated temperature (*Aspergillus, Cladosporium, Thamnidium* sp). Yeasts also grow and involve in spoilage of food under appropriate conditions. These grow over psychrophilic and mesophilic temperature range.

Storage and Spoilage

Temperature of storage is most important parameter affecting the spoilage of highly perishable foods (Ex. fish).

Nature of food need to be considered while selecting storage temperature. Maintaining all foods at refrigerated temperature is not advisable as it affects the quality of food.

Impact of Time

Time is a critical factor when considering growth rates of microbial pathogens, in addition to temperature, as it relates to a product's shelf life.When time alone at ambient temperatures is used as a control product safety, the duration should be equal to or less than the lag phase of the pathogen(s) of concern in the product. For refrigerated food products, the shelf life required for safety may vary depending on the storage temperature. For instance, lag time for growth of *L. monocytogenes* at 10°C is 1.5 d, while at 1°C it is ~3.3 d. The generation time is 5-8 h at 10°C and 62-131 h at 1°C. According to the USFDA Pathogen Micromodel, a temperature shift from 10 to 25°C decreases the lag time of *L. monocytogenes* from 60 to 10 h. In a similar manner, a pH increase from 4.5 to 6.5 decreases the lag time from 60 to 5 h. So, the safety of a product during its shelf life differs, depending upon other conditions such as temperature of storage, pH of the product, *etc.*

Storage/Holding Conditions

The storage conditions include the storage/holding temperature, the time/temperature involved in cooling of cooked items, and the relative humidity to which the food or packaging material is exposed. Other factors involved are the effectiveness of the packaging material. Foods that are cooked or re-heated and served or held hot may require appropriate time/temperature control for safety. For instance, the primary organism of concern for cooked fish products is *Listeria monocytogenes.* Outbreaks of food borne illness had resulted due to slow cooling of food, as it permits growth of pathogenic bacteria, particularly the spore-forming pathogens that have relatively short lag times and the ability to grow rapidly. For instance, *C. perfringens*, and *Bacillus cereus.*

The effect of the relative humidity of the storage environment on the safety of foods is more nebulous. The relative humidity may or may not alter the a_w of the food. Foods that depend on a certain a_w for safety or shelf life considerations need to be stored such that the environment that does not markedly change this characteristic.

iii. Relative Humidity (RH)

Relative humidity of the food storage environment refers to percentage of moisture present in the atmosphere. RH and water activity are closely related and RH is measure of aw of the gas phase of atmosphere.

$$RH = 100 \times a_w$$

Effect of RH on Food being Stored

RH of storage environment influences aw of foods and growth of microorganisms on food surface. Foods of low a_w stored in high RH environment leads to transfer of water from environment to the food, increasing aw of food until equilibrium is reached. Condensation of moisture on food surface results in localized regions of high a_w on surface and subsurface. This favors growth of microorganisms which were viable but unable to grow due to low a_w, increases a_w of immediate environment due to metabolic activity of microorganisms which produce water as end product of respiration. Thus microorganisms grow and cause spoilage of food which was considered safe microbiologically due to the growth of microorganisms requiring high a_w. Foods of high a_w when stored in low RH environment lose moisture, become flaccid and unfit for consumption due to loss of quality.

RH and Spoilage of Food

1. RH and temperature of environment are related, and as the temperature increases RH decreases.
2. Foods that are susceptible for spoilage by yeasts, molds and certain bacteria should be stored under low RH conditions.
3. Improperly wrapped animal meat kept in refrigerator (high RH) undergone surface spoilage.
4. Selection of suitable RH condition for storage is necessary to avoid surface microbial growth and maintain desirable qualities of food as food may lose/take up moisture under improper RH condition and lose its quality.

Chapter 3

Microorganisms in Fish and Fishery Products

3.1. Microbial Spoilage of Fresh Fish

The foods of plant and animal origin contain several microorganisms and these along with natural food enzymes become active soon after the death of animal and harvest of plant foods. The intrinsic characters of the food and the environmental conditions influence the type and extent of microbial activity leading to spoilage which is characterized by deteriorative changes in food quality making it unfit for human consumption. The different preservation techniques have mainly aimed at reducing or eliminating microorganisms in food thereby prolonging or preventing the spoilage of foods. The preservation of foods is commonly done with altering intrinsic and extrinsic factor by use of various process such as low temperature, high temperature, drying, radiation and chemicals.

Bacterial spoilage is caused by the activities of microorganism associated with the fish. Bacterial spoilage of fish begins only after the completion of rigor mortis, which results in the release of products of protein denaturation due to decrease in pH, which is utilizable by bacteria. Thus, prolonging rigor mortis helps to delay spoilage and thereby keeps fish fresh.

Spoilage of both marine and fresh water fish occurs in the same manner. Fish contain high levels of protein and non- protein nitrogenous constituents (16~20 per cent), lack carbohydrate, and have varying amounts of fat depending on the species of fish. The non-protein nitrogenous compounds in fish include free aminoacids, volatile nitrogen bases-ammonia and trimethyl amine (TMA), creatine, taurine, betaines, uric acid, anserine, carnosine and histamine. Spoilage of fish begins from the surface, gill and intestine because of high bacterial load. From gills, intestine and surface microorganisms gradually migrate to adjacent tissue and cause spoilage. Spoilage organism first utilizes simpler compounds and later fish protein releasing various off-odour compounds.

The microbial laod and and types of microorganisms associated with the freshly harvested fish directly reflects on the microflora of the immediate environment from which fish is harvested. Further, several microorganisms are added to the fish from harvesting gear and net, from the boat deck, fish contact surface, fish holds and fishermen. Once the fish dies, microorganisms present on the body surface, gills and the intestine start multiplying and increase in numbers causing deteriorative changes in fish. As the time lapses the spoilage proceeds faster making the fish unfit for human consumption. The spoiling fish is generally characterized by the loss of bright body coloration, fading of gill colour, sunken eyes and development of off odour metabolites of spoilage bacteria. The spoilage flora of fresh fish is generally dominated by the Gram negative bacteria. The effective way to prevent or delay spoilage of fresh fish is by reducing the activity of spoilage microorganisms which can be achieved by lowering the temperature of holding the harvested fish.

3.1.1. Microbiology of Finfish

The subsurface flesh of live, healthy fish is considered sterile and should not present any bacteria or other microorganisms. On the contrary, as with other vertebrates, microorganisms colonise the skin, gills and the gastrointestinal tract of fish. The number and diversity of microbes associated with fish depend on the geographical location, the season and the method of harvest. In fact, standard culture-dependent methods can only recover between 1 to 10 per cent of total bacteria present in any given sample. More accurate, molecular-based methods have not yet been used to address this issue. Gastrointestinal tracts and gills typically yield high bacteria numbers, although these are influenced by water quality and feed. Microorganisms are found on skin, gills and in the intestines of live and newly caught fish. The total number of bacteria ranges from 10^2 to 10^7 cfu/cm^2 on the skin surface and between 10^3 and 10^9 cfu/g for both in the gills and intestines. Fish harvested from clean and cold waters will present lower bacterial numbers than fish from eutrophic and/or warm waters. Also, fish from polluted warm waters contain up to 10^7 cfu/cm^2. However, potential human pathogens may be present in both scenarios.

The autochthonous bacterial flora of fish is dominated by Gram-negative genera including: *Acinetobacter, Flavobacterium, Moraxella, Shewanella* and *Pseudomonas*. Members of the families Vibrionaceae (*Vibrio* and *Photobacterium*) and the Aeromonadaceae (*Aeromonas* spp.) are also commonaquatic bacteria, and typical of the fish flora. Gram-positive organisms such as *Bacillus, Micrococcus, Clostridium, Lactobacillus* and *Coryneforms* can also befound in varying proportions. It is crucial to mimic the environmental physico-chemical parameters when isolating bacteria from fish. For example, some species (most Vibrios) require sodium chloride for growth; whenever possible; several culture media containing sodium chloride and more than one incubation temperature should be used. It must be noted that mesophilic bacteria can rapidly overgrow psychrotropic organisms.

It is apparent from the above that there is potentially a very diverse range of organisms present on fish. However, numbers of pathogenic bacteria in raw fish tend to be low, and risk associated with the consumption of seafood is low. In addition, during storage indigenous spoilage bacteria tend to outgrow potential pathogenic bacteria. Shelf life depends on the initial microflora on the fish, potential contaminants added during handling and processing, and conditions of storage.

Recognised Specific Spoilage Organisms (SSOs)

The degree of spoilage leading to sensory rejection of fish is partly dependent on the perception of the consumer. Not all the bacteria growing on a food will lead to the production of objectionable characteristics; a minority are often associated with the majority of the spoilage. The concept of specific spoilage organisms (SSOs) is not new; yoghurt spoilage by yeasts and clostridial spoilage of cheese are examples where it has been recognised for many years that a particular minority of the microbial flora present in the product is responsible for its spoilage. For fresh fish, realisation that the bulk of the microbial population on newly caught fish does not cause off-flavours and off-odours stems from work. The spoilage of a product may be strongly influenced by the conditions under which the product is held; therefore, the characterisation of spoilage of each product must be made before the identification of the responsible agent(s) can proceed. Bacteria identified as being associated with the spoilage process of fresh fish are as follows:

Pseudomonas spp.

The Pseudomonadaceae family represent a large and poorly defined group of microorganisms. They are generally characterised as Gram-negative rods, motile with polar flagella, oxidase-positive, catalase-positive, obligate respiratory bacteria. The spoilage compounds associated with the growth of psychrotrophic *Pseudomonas* spp. on fish are diverse and in many cases species-specific. *Pseudomonas* spp. mediated spoilage is characterised by 'fruity', 'oniony' and'faecal' odours from the production of ketones, aldehydes, esters and non-hydrogen sulphide sulphur-containing compounds such as methyl sulphide. Members of the genus are able to produce pigments, and proteolytic and lipolytic enzymes that may affect the quality of fresh and, more especially, processed (*e.g.* frozen) fish products. The spoilage of freshwater fish is generally ascribed to the growth of *Pseudomonas* spp. and they are considered SSO of iced freshwater fish.

Shewanella putrefaciens

Shewanella putrefaciens is considered an SSO of temperate-water marine fish species stored in ice; it is often isolated as about 1-10 per cent of the total flora of fresh fish from temperate marine waters. It is also present in fresh water, and may play some role in the spoilage of freshwater fish. It is able to grow as fast as, or faster than the rest of the flora of ice-stored fresh marine fish.

Shewanella putrefaciens spoilage of fish is due to its biochemical action on muscle, *i.e.* its ability to reduce TMAO to TMA, produce hydrogen sulphide (H_2S) from cysteine, form methylmercaptane (CH_3SH) and dimethylsuphide ($(CH_3)_2S$) from methionine and produce hypoxanthine (Hx) from inosine monophosphate (IMP) or inosine, plus other characteristic compounds of the species responsible for spoilage. It is thus an important spoilage organism of gadoid fish such as cod, for which the most compelling evidence showing spoilage as a result of Shewanellaputrefaciens growth has been collected. It is also able to produce hydrogen sulphide and a range of other off-odour compounds.

In terms of its taxonomy, *Shewanella putrefaciens* strains are characterised as microaerophiles or anaerobes, they are heterogeneous, and recent taxonomic description using modern molecular methods bifurcated this species, with several new species being described. Initially, mesophilic strains of *Shewanella putrefaciens* were identified as *Shewanella algae, Shewanella waksmanii, Shewanella affinis,* and *Shewanella aquimarina*. Later, additional psychrotrophic species such as *Shewanalla baltica, Shewanella oneidensis, Shewanella gelidimarina, Shewanella frigidimarina, Shewanella livingstonensis, Shewanella olleyana, Shewanella denitrificans,* and *Shewanella profunda* were described. In fact, *Shewanella baltica* has been identified as the main H_2S producer in cod during cold storing.

Photobacterium phosphoreum

Recognised for some time as being present on spoiling fish, *Photobacterium phosphoreum* increased in notoriety when it was proposed that reduction of TMAO to TMA limited the shelf life of MAP cod fillets. Because no other TMAO-reducing bacteria were present in sufficient numbers to produce the quantities of TMA that were related to rejection, it was proposed that this organism, owing to its cell size and activity, was capable of being in a significant numerical minority on the MAP fish but still able to yield the majority of the TMA thought to be responsible for the rejection.

Ph. phosphoreum has also been identified as responsible for histamine fishpoisoning. This type of intoxication occurs when bacteria convert the histidinepresent in fish into histamine. Some *Ph. phosphoreum* strains have great capacity as histamine producers even under refrigeration conditions.

Brochothrix thermosphacta and Lactic Acid Bacteria

B. thermosphacta is a well-characterised psychrophilic spoilage organism of meat. Growing evidence suggests a role for *B. thermosphacta* in the spoilage of some MAP fish. Recent studies have investigated the dominance of *B. thermosphacta* on spoiling fish in a 40 per cent CO_2/30 per cent N_2/30 per cent O_2 MAP. Acetate production has been reported as a good indicator of spoilage by this organism.

MAP studies have also demonstrated this organism's sensitivity to oxygen and have shown that they are also inhibited by high CO_2 concentration.

Pathogens

Human pathogenic bacteria can be part of the initial microflora of fish, posing a concern for seafoodborne illnesses. These pathogens can be divided into two groups: organisms naturally present on fish (Table 3.1); and those that although not autochthonous to the aquatic environment, are present there as result of contamination (anthropomorphic origin or other) or are introduced to the fish during harvest, processing or storage (Table 3.2).

Table 3.1: Indigenous Bacterial Pathogens Typically Present in Fish

Organism	*Temperature Range*	*Estimated Minimum Infective Dose*
Clostridium botulinum-non-proteolytic type E	3-26°C	00.1-1 μg toxin lethal dose
Pathogenic Vibrio spp.	10-37°C	High for most species,
Vibrio cholera		10^5-10^6 cells/g
Vibrio parahaemolyticus		(exception: *V. vulnificus*)
Vibrio vulnificus		
other *vibrios*		
Aeromonas spp.	5-35°C	Unknown
Plesiomonas shigelloides	8-37°C	Unknown

Table 3.2: Non-indigenous Bacterial Pathogens Frequently Present in Fish

Organism	*Primary Habitat*	*Minimum Infective Dose*
*Listeria monocytogenes**	Soil, birds, sewage, stream water, estuarine environments, and mud	Variable depending on the strain (>10^2 cells/g)
Staphylococcus aureus	Ubiquitous, human origin	10^5-10^2 cells/g. Toxin levels 0.14 - 0.19 μg/kg body weight
Salmonella spp.	Intestinal tract of terrestrial vertebrates	from < 10^2- >10^2
Shigella spp.	Human origin	10^1-10^2
Escherichia coli	Fecal contamination	10^1-10^3
Yersinia enterocolitica	Ubiquitous in environment	High (10^7-10^9 cells/g)

* Some sources considered *L. monocytogenes* as part of the natural aquatic flora.

Endogenous chemicals, algal toxins, human viruses, bacteria and higher parasites all present some risk associated with fish consumption. Of these, it is bacterial risks that increase after capture of the fish.

There are few human bacterial pathogens that can cause primary infections or disease and are capable of persisting in the aquatic environment. Fewer still are capable of growing on fresh fish. The remaining few bacteria present a major risk involved with the consumption of raw seafood such as sushi or oysters, but with proper cooking, these risks are substantially reduced.

3.1.2. Microbiology of Mollusc

The phylum Mollusca, which includes oysters, clams, mussels, cockles, scallops, squid, octopi, chitons, snails, slugs, whelks, abalones, and tooth shells are a group of soft-bodied animals that usually secrete external protective shells.

Bivalves feed by filtering large volumes of water across their gills to obtain oxygen and food. In the process of filtering water, they also trap bacteria, viruses, chemical contaminants and other impurities on the mucus of the gills. Many microorganisms ingested by shellfish survive the digestive process.

The term shellfish in this context is limited to oysters, clams, mussels, cockles and scallops. Other species of shellfish such as crabs, lobster, and shrimp, while commercially valuable, are not as vulnerable to contamination through pollution of their beds as the bivalves.

Microflora

The microflora of molluscan shellfish directly reflects the environment from which the shellfish is harvested. The initial microflora of molluscan shellfish is comprised of the natural commensal microorganisms and the microorganisms accumulated from the water during feeding. Because molluscan shellfish are filter feeders, they accumulate pathogenic microorganisms from polluted waters. Whatever contaminants are in the water will eventually get into the shellfish. The microflora are dependent upon a number of factors such as season, and the environment, which affects the temperature, salinity, pH, nutrient concentration and pollutants of the water column. Raw and partially cooked shellfish, mostly in oysters, clams and mussels, are vectors of disease transmission. The following pathogenic microorganisms causing illness in humans are characteristically associated with molluscan shellfish:

Bacteria

Vibrio parahaemolyticus | *Vibrio vulnificus*
Vibrio cholerae 01 | *Vibrio cholerae non-01*
Vibrio fluvialis | *Vibrio mimicus*
Vibrio hollisae | *Vibrio furnissii*
Aeromonas hydrophila group | *Plesiomonas shigelloides*
Staphylococcus aureus | *Salmonella* spp.
Shigella spp. | *Campylobacter jejuni*
Campylobacter coli | *Escherichia coli*
Bacillus cereus

Virus

Hepatitis A	
Hepatitis non-A, non-B	Norwalk virus
Snow Mountain virus	Other viral agents

A variety of commensal microfloras are found in molluscan shellfish. The microflora of molluscan shellfish at harvest consists predominantly of Gram-negative rods while the Gram-positives constitute a minor portion of the flora. The largest groups found are the *Pseudomonads, Vibrios,* and *Aeromonads.* Other microorganisms found in lesser numbers are the *Achromobacter* spp., *Flavobacterium* spp., *Pseudomonas* spp., *Acinetobacter* spp., *Micrococcus* spp., *Enterococci* spp., *Bacillus* spp., *Alcaligenes* spp., *Moraxella* spp., and yeast cellsof Rhodotorula rubra and *Trichosporon* spp. At harvest, molluscan shellfish have an Aerobic Plate Count (APC) of approximately 1000 to 100,000/g.There are microorganisms, observed microscopically, that have never been cultured on laboratory media. They are termed viable non-culturable microorganisms. For example, spirochetes of the genus Cristispira and Saprospira, have been observed microscopically in the digestive tract of Eastern and Pacific oysters, but have never been cultured.

Shellfish may contain higher levels of pathogens than are found in the water in which they grow. The safety of shellfish is predicated by the cleanliness of the growing area waters from which they are harvested, and the sanitary practices applied during harvesting and shipping. The most important factor in controlling bacterial growth in shellstock is temperature. Shellstock should be refrigerated at temperatures at or below 10°C (50 °F). At this temperature the multiplication of faecal coliforms (*E. coli, Klebsiella, Enterobacter,* and *Citrobacter* species) and Vibrios, including *V. vulnificus,* areprevented, with the exception of *A. hydrophila.* Water used for shellstock washing should be of good sanitary quality, to avoid possible contamination of the shellstock. The washing of sediment and detritus soon after harvesting will prevent further contamination.

Spoilage

Achromobacter, Pseudomonas, Flavobacterium, Micrococcus, Proteus, Alcaligenes, and *Pseudomonas fluorescens* are the responsible agents of spoilage in oysters held atvarious temperatures. Shellfish, particularly shucked meats, are an excellent medium for growth of bacteria. The initial flora found in shellfish is the major factor of spoilage in bivalves. The environment has an effect on the bacterial population found in shellfish. These organisms may play a significant role in the spoilage processes. The spoilage process in shellfish may be divided into three stages: increase of acidity, abundant gas production, and proteolysis. The high incidence of proteolytic bacteria and other types capable of fermenting glucose in the natural flora may be responsible in post-mortem spoilage of shellfish. Glycogen, a carbohydrate, is stored in the bivalve as a source of energy. Glycogen is hydrolysed by the initial flora found in shellfish, thus producing acid and gas. This process changes the bacterial profile to microorganisms favouring low pH; these include *lactobacilli, streptococci,* and yeasts. *Lactobacilli* seem to be the predominant microflora in shellfish, particularly oysters stored at 7 °C, whereas bacterium such as *Pseudomonas, Achromobacter* and *Flavobacterium* levels decrease during storage.

Pathogens

These organisms can be present in the shellfish from the point of harvesting, during processing, and to the point of sale. They can be introduced to the shellfish from the environment or added during processing. The control of growth and survival of these organisms can be reduced by the use of good manufacturing practices including immediate and proper refrigeration, prevention of cross-contamination, and harvesting of shellfish in the cooler months.

Vibrios are Gram-negative facultative rods naturally occurring in the marine water (estuaries, coastal areas), primarily in brackish or saltwater. Diseases caused by Vibrios are enteric in nature, ranging from epidemic cholera to sporadic cases of diarrhoea.

V. parahaemolyticus causes gastroenteritis lasting 24 - 48 hours with abdominalpain, diarrhoea, nausea, headache and fever. It is a common marine bacterium, widespread in both polluted and unpolluted waters. Major outbreaks are associated with the warmer months. *V. parahaemolyticus* multiplies rapidly at temperatures above 20°C. The symptoms of *V. vulnificus* are fever, chills, and nausea within 24 - 48 hours of onset. Death has been reported to occur within 36 hours from onset. It is one of the most severe foodborne infectious diseases, with a fatality rate of 50 per cent for individuals with septicaemia. Oysters are the most hazardous food associated with *V. vulnificus*. The bacterium is widespread in estuarine waters particularly in warm waters of >20°C. The occurrence of *V. vulnificus* causing septicaemia and death has been associated with the consumption of oysters. Control of *V. vulnificus* is by rapid refrigeration during warm weather months. Hazards from this bacterium can be controlled by thorough cooking of shellfish, and preventing cross contamination once the product is cooked. Illness caused by *V. cholerae* non-01 is generally a less severe gastroenteritis than that caused by *V. cholerae*. It causes diarrhoea, abdominal cramps, fever, nausea, and vomiting. Onset of symptoms occurs in 48 hours and can be severe, lasting 6 - 7 days.

Vibrio cholerae 01 and 0139 causeing epidemic cholera, vary from mild watery diarrhoea to acute diarrhoea with characteristic rice water stools. *V. cholerae* 01 is transmitted by faecal contamination of food or water. The control of *V. cholerae* 01 in molluscan shellfish, particularly in oysters, is the same as previously discussed for the non-01 strains. In addition, processors should know the product source, *e.g.* imported product from a country experiencing an epidemic. There is several other *Vibrio* spp. associated with shellfish-borne illness outbreaks, including *V. fluvialis, V. mimicus,* and *V. hollisae*. The frequency of infection of these organisms is no different from other Vibrios; however, the pathogenic severity is lower than *V. cholerae* 01, *V. parahaemolyticus* and *V. vulnificus.*

C. jejuni requires reduced levels of oxygen (3 - 5 per cent) and 2 - 10 per cent carbon dioxide for optimal growth conditions. It is considered a fragile and sensitive organism to environmental stresses such as oxygen, drying, low pH, salinity, acidity and heat. It causes diarrhoea, abdominal pain, fever, and nausea. Onset of illness is from 3 to 5 days after ingestion of contaminated food, and illness lasts 7 - 10 days.

Published Microbiological Criteria

The published microbiological criteria below are included as a guide, and where indicated are intended to be used for official control purposes.

Test	*Acceptable Levels in Molluscan Shellfish*
Standard Plate Counts	<500,000 cfu/g
E. coli	<230/100 g
Enterotoxigenic *E. coli*	<1000 per gram, negative for heat-labile toxin (LT)
Salmonella	or heat-stable enterotoxin (ST) Negative for the presence Negative for
V. cholerae	the presence of toxin producing 01 or non-01 organisms <10,000 MPN
V. parahaemolyticus	per gram
Staph. aureus	Negative for staphylococcal enterotoxin or when the viable MPN count
C. jejuni	is <10,000 negative for the presence
V. vulnificus	Under review

3.1.3. Microbiology of Crustaceans

Crustaceans are invertebrate animals and represent a class of the phylum Arthropoda. They possess a segmented body, jointed limbs and a chitinous exoskeleton, and are usually aquatic, *e.g.* crabs, lobsters, crayfish and shrimp. Crayfish, lobsters and crabs have ten legs; the front pair ends in claws. Shrimps, prawns and scampi are temperate species commonly eaten by man.

Initial Microflora

As with finfish, the level and type of the initial microflora of crustacean shellfish will reflect a combination of factors, which include the environment from which they have been captured or harvested; their feeding and living habits; the geography of the area they are captured from or are grown in; the season; and the temperature and quality of the waters in which they exist. Following capture or harvest, the flora will change depending upon the methods of handling and/ or the environmental conditions to which they are exposed. Microbiological risks associated with crustacean shellfish increase in proportion to the degree of handling that the product undergoes, particularly after it has been cooked. Control measures to prevent these potential hazards occurring include training and supervision of operatives as well as Good Hygienic Practices (GHP). Poor hygienic procedures will increase the probability of contamination with bacteria of public health significance. The structure of the circulatory system of crabs is not closed, which means that the haemolymph can be a reservoir of bacteria, particularly for members of the genus Vibrio.

The microbial flora of crustaceans from temperate waters differs from that of crustaceans from warmer, tropical waters. Spoilage bacteria associated with temperate crustaceans include *Pseudomonas* spp., *Shewanella putrefaciens* and members of the Acinetobacter-Moraxella group. In contrast, the genera associated with tropical species include Enterobacteriaceae, *Vibrio* spp., *Pseudomonas* spp., Alcaligenes, Corynebacteria, *Micrococcus* spp., and *Achromobacter* spp. Hence, from the time of capture and processing, bacterial contamination and spoilage are inevitable.

As with finfish, there is evidence that the shelf life of tropical crustaceans is longer than that of temperate-water species stored correctly in ice - for example, up to 16 days for tropical species compared with 8 - 10 days for temperate water species. The psychrotrophic bacteria responsible for causing spoilage are not present on freshly captured crustaceans from tropical waters in such high levels as are found in crustaceans from cooler waters and, thus, the rate of deterioration of the crustacean shellfish by these organisms is reduced, especially if they are stored in ice or iced water.

Spoilage

The main genera of Gram-negative bacteria responsible for the spoilage of crustacean shellfish are the same as those that cause spoilage in finfish; they include *Sh. putrefaciens* and *Pseudomonas* spp.

Prawns and Shrimps

Freshness indicators for prawns and shrimps are for the shells to be crisp and dry and feel sharp and cool to the touch, and have a sweetish, slightly iodine smell. In comparison, spoiling prawns and shrimps have shells that are soft, wet and 'soapy' or 'jammy' to the touch. Heat is produced if the shrimps decompose in a confined space, and a strong ammoniacal odour is produced, indicative of protein degradation and the growth of the principal spoilage bacterium *Sh. putrefaciens.*

Crabs and Lobsters

Crabs and lobsters should be alive when boiled or sold. If boiled, the tightness of limbs will indicate whether they were killed by immersion in boiling water or, for crabs, immediately prior to boiling. The shells have a crisp, bright and dry appearance, whilst boiled crustaceans have a pleasant slightly sweet smell. With spoiled crabs and lobsters, there is no indication of life. If the tail drops, the joint between the tail and carapace is open and the limbs hang loosely. The shell will be soft, dull and sticky, with visible decomposition or green discoloration seen through the membrane exposed by the dropping tail. An offensive odour is also produced.

Pathogens

Salmonella spp.

The genus *Salmonella* contains a wide variety of 'species' pathogenic for man or animals, and usually for both. They are mesophilic microorganisms usually residing in the intestinal tracts of human beings and warm-blooded animals. Hence, *Salmonella* species are more commonly found in crustaceans reared by aquaculture or growing in faecally contaminated brackish water compared with species of marine origin. Moreover, animal manure, a common pond fertiliser used in aquaculture, often contains *Salmonella* and other enteric pathogens and has been demonstrated to contaminate crustaceans growing in these environments. Aquatic birds and the hands of workers have also been identified as sources of the pathogen. *Salmonella* cannot be completely removed from raw shrimp during processing, even if GMP is followed, although efficient cooking will quickly eliminate it.

Listeria spp.

The presence of *L. monocytogenes* in cooked ready-to-eat crustaceans is recognised as a potential risk to selected groups of individuals: the very young and old, pregnant women and the immuno-compromised. To date, no incidence of foodborne illness caused by *L. monocytogenes* has been attributed to crustaceans. However, the pathogen can grow at temperatures as low as 3°C.

Staphylococcus aureus

Staph. aureus is not a member of the intrinsic flora associated with crustacean shellfish. The main source of contamination is the hands of processing personnel; about 50 - 70 per cent of healthy individuals carry toxin-producing strains of *Staph. aureus* on their hands, as part of the natural bacterial flora, whilst boils and cuts are also common sources of the pathogen.

Crustaceans have been linked with food intoxications attributed to *Staph. aureus* because of the ubiquity of this organism in the processing environment, combined with the degree of handling of these types of product, particularly crabmeat.

Clostridium botulinum

C. botulinum type E is commonly associated with fish and fishery products,although contamination by types A and B may occur from the environment. The pathogen has resistant spores, so is able to survive heat treatments that eliminate vegetative cells, *e.g.* pasteurisation and cooking. Once the competing microflora has been removed by a heat treatment, GMP should be followed in the processing plant to ensure that the final product is held at 3°C, especially in products that have been vacuum packaged.

Researchers have demonstrated that inoculated shrimp supported toxin production by *C. botulinum* type E at 10°C but not at 4°C.

Vibrio spp.

Vibrio cholerae, Vibrio parahaemolyticus and *Vibrio vulnificus* are species that have been reported to be associated with raw crustaceans harvested from estuarine waters as well as from undercooked crustacea when harvested from contaminated waters. Crabs are covered in a chitin exoskeleton, which can become colonised by *V. cholerae*; therefore, it is important that the cooking stage is controlled to eliminate the pathogen from the product and that procedures are in place to prevent any cross-contamination of the cooked product.

V. cholerae is divided into two groups, *V. cholerae* O1 and *V. cholerae* non-O1. The former, the causative agent of cholera, is widely distributed in aquatic environments, although more recently the non-O1 type has also been demonstrated to cause foodborne illness in crustaceans. *V. parahaemolyticus*, a marine halophile, occurs in crustaceans from warm environments. It is more commonly implicated in countries where crustacean shellfish are eaten raw, *e.g.* Japan, South America.

V. vulnificus is an invasive and lethal pathogen associated with wound infections, but has the potential to cause fatal foodborne illness. The organism is widespread in estuarine waters and is a common isolate from harbour water in warm climates. Contamination by this organism can be controlled by rapid chilling of crustaceans after harvesting and avoidance of time/temperature abuse during distribution and processing chains.

Campylobacter jejuni and Campylobacter coli

C. jejuni has not, to date, been implicated in a case of foodborne infection associated with crustaceans. However, the potential hazard exists as *C. jejuni* was isolated from 36 (15 per cent) of 240 samples of freshly hand-picked blue crab (*Callinectes sapidus*) meat, whilst 5.8 per cent of samples contained *C. coli*. Quantitative levels were below limits of detection in all cases (<0.30 MPN/g) (27).

Aeromonas hydrophila and Aeromonas sobria

A. hydrophila and *A. sobria* are known to be fish pathogens and are commonly isolated from ponds used for aquaculture. However, they have not been confirmed to be causal agents of foodborne illness associated with crustaceans, although *A. hydrophila* has been isolated from prawns. *A. hydrophila* is able to grow well at refrigeration temperatures (0 - 2°C) so that GHP coupled with controlled low temperature storage is essential to prevent the pathogen from reaching levels that may present a hazard.

Parasites

The use of animal excreta in aquaculture has also been associated with foodborne parasitic infections of crustaceans, particularly trematodes. Parasitic infections are controlled by either cooking infected material prior to consumption, or freezing at a temperature of -23°C for 7 days.

Viruses

Crustaceans grown in waters contaminated with sewage may become carriers of viruses such as Hepatitis A, Norovirus, Caliciviruses, Astroviruses and non-A, non-B Hepatitis viruses. Of these, Hepatitis A virus presents the most serious hazard; Both Hepatitis A virus and Norovirus are stable pathogens and are relatively resistant to inactivation. They can survive low pH levels, consistent with that in the stomach, or chlorine concentrations similar to those found in drinking water. Norwalk viruses can also withstand heating to 60°C.

3.2. Mechanism of Microbial Spoilage

Microbes decompose and spoil fish in ways as given below:

- Utilisation of readily utilisable simple substances such as non-protein nitrogenous (NPN) substances
- Hydrolysis of complex tissue components (*e.g.* proteins) into simpler substances (*e.g.* amino acids) and their subsequent utilisation
- Breakdown of nucleotides

3.2.1. Utilisation of Readily Utilisable Simple Substances

The NPN compounds are readily utilisable simple substances consisting of TMAO, Urea and free amino acids.

i. Trimethylamine Oxide (TMAO) Degradation

TMAO is an osmoregulatory substance present in the tissue of marine finfish and shellfish. Negligible quantities of TMAO is present in freshwater fishes. The elasmobranchs contains the highest and the flat fishes the lowest TMAO. The muscle of sharks contains 750-1480 mg per cent TMAO. White fleshed demersal fishes contains more TMAO than dark fleshed pelagic fishes.

TMAO is subsequently degraded through reduction to trimethyamine, dimethyl amine, methyl amine and finally to ammonia. Reduction of TMAO involves reduction of one methyl group in each step with its liberation as methane. TMAO is reduced to TMA enzymatically by the enzyme TMAO reductase. This enzymatic reduction takes place by microbial enzyme as well as by tissue enzymes. The TMAO reduction is mainly associated with the genera of bacteria such as *Alteromonas, Photobacterium, Vibrio, Shewanella putrefacians, Aeromonas* and intestinal bacteria of Enterobacteriaceae.

TMA is responsible for the typical fishy odour of the marine teleosts. The content of TMA in fish tissue is expressed as TMA-N and is used as an indicator of microbial spoilage. The level of TMA found in fresh fish rejected by sensory panels varies between fish species, and is around 10-15 mg N/100g.

In fresh water fishes that lacks TMAO, ammonia is produced by deamination of the amino acids by microbes. The compounds produced due to reduction of TMAO such as TMA, DMA, MA and ammonia are volatile and produce off odour of spoiling fish. These compounds are collectively called as "volatile base nitrogen". TVBN is also used as an indicator of microbial spoilage of fish. TVBN is a better indicator of spoilage than TMA because it includes intermediate products and the end products. The level of TVB-N in freshly caught fish is between 5 and 20 mg N/100g and the limit of acceptability of ice stored fish is 30-35 mg N/100g.

During chilled or frozen storage of fish, when bacterial growth is inhibited, the reaction of conversion of TMAO into TMA is replaced by slow conversion of TMAO into DMA and formaldehyde. The formation of these products may cause severe quality changes such as increased denaturation, changes in texture and loss of water binding capacity or spoilage during prolonged storage.

Urea Degradation

Urea is an osmo- regulatory substance present in the tissue of cartilaginous fishes like sharks, skates and rays. It is decomposed to carbon di oxide and ammonia by the microbial enzyme, urease. This process benefits the bacteria as they use the ammonia for respiration. The production of ammonia also raises the pH of the substrate which promotes the growth of many urea degrading bacteria inhibiting competition by many other species. The urea degradation is associated with bacteria such as *Helicobacter pyroli, Klebsiella pneumoniae, Proteus* spp. and *Micrococcus luteus.*

Free Amino Acid Degradation

The free amino acids are present in the fish tissue in small quantities. After the death of fish, due to autolysis, free amino acids are formed from tissue proteins resulting in nutrient rich medium for microbial activity. They are also formed from hydrolysis of tissue proteins by proteolytic enzymes secreted by microbes such as *Pseudomonas, Sarcina* when the available free amino acids are scanty or exhausted. The free amino acids are absorbed into the microbial cell where they are metabolised through decarboxylation and deamination by specific enzyme system of microbes. The metabolic end products come out of the microbial cell. Degradation of free amino acids takes place in three ways:

a. Deamination

Deamination is the process of the removal of amino group from the amino acids and the amino group isthen reduced to ammonia. The microorganisms associated are Pseudomonas, Enterobacteriaceae and Lactic acid bacteria. Deamination occurs in four ways.

- Oxidative deamination: It takes place in the presence of oxygen with production of keto acid and ammonia. The keto acid eventually turns into aldehydes, lower fatty acids and carbon di oxide.
- Reductive deamination: It occurs in the presence of hydrogen producing a saturated fatty acid and ammonia.

- Hydrolytic deamination: In this process, free amino acids are hydrolysed in the presence of water with the production of hydroxyl acid and ammonia. The hydroxyl acid is decarboxylated with the liberation of carbon di oxide and production of alcohols. The alcohol is oxidised further to aldehydes, ketones and lower fatty acids.
- Desaturative deamination: The amino acid is converted to unsaturated fatty acid with the formation of ammonia.

b. Decarboxylation

Decarboxylation is the removal of the carboxylic group (-COOH) from the amino acids with the production of carbon di oxide and amines. These amines are called biogenic amines and include histamine, cadaverine, putrescine, agmatine *etc.* These biogenic amines are biologically active low molecular weight nitrogenous compounds. The biogenic amines formed from amino acids and the respective enzymes are given in Table 3.3.

Table 3.3: Formation of Biogenic Amine from Amino Acids

Amino Acid Precursor	*Aminoacid Decarboxylase*	*Biogenic Amines*
Histidine	Histidine decarboxylase	Histamine
Tyrosine	Tyrosine decarboxylase	Tyramine
Tryptophan	Tryptophan decarboxylase	Tryptamine
Lysine	Lysine decarboxylase	Cadaverine
Phenylalanine	Phenylalanine decarboxylase	Phenylethylamine
Arginine	Arginine decarboxylase	Agmatine
Ornithine	Ornithine decarboxylase	Putrescine ↕ ↔ Spermine; Spermidine ↔ Spermine

c. Deamination and Decarboxylation

Here, both the amino group (-NH_2) and carboxylic group (-COOH) are removed from the amino acids. It occurs in three ways *viz.* (i) oxidative deamination and decarboxylation, (ii) reductive deamination and decarboxylation and (iii) hydrolytic deamination and decarboxylation.

3.2.2. Formation of Volatile Sulphur Compounds

Volatile sulphur compounds are typical components of spoiling fish and most bacteria identified as specific spoilage bacteria produces one or other volatile sulphides. *S. putrefaciens* and some Vibrionaceae produce H_2S from sulphur containing amino acid, l-cysteine. Alteromonas and Pseudomonas are active sulphide producers. Methylmercaptan (CH_3SH) and dimethylsulphide ($(CH_3)_2S$) are formed from methionine (Table 3.4). The volatile sulphur compounds are foul smelling and can be detected at very minute levels.

Table 3.4: Typical Spoilage Compounds Formed during Spoilage of Fresh Fish

Specific Spoilage Organism	*Typical Spoilage Compounds*
Shewanella putrefaciens	TMA, H_2S, CH_3SH, $(CH_3)_2S$, Hx
Photobacterium phosphoreum	TMA, Hx
Pseudomonas spp.	Ketones, aldehydes, esters, non- H_2S sulphides
Vibrio naceae	TMA, H_2S
Anaerobic spoilers	NH_3, acetic, butyric and propionic acid

3.2.3. Hydrolysis of Complex Tissue Components into Simpler Substances and their Utilisation

When the simple substances such as TMAO, urea and free amino acids are completely exhausted, the microbes starts hydrolysing complex tissue components such as proteins and lipids to produce simpler substances. For utilising protein, proteolytic group of bacteria secrete proteolytic enzymes from their cells to the surroundings. These enzymes hydrolyse protein into amino acids. These amino acids enters the microbial cell and are metabolised and the end products are liberated out of cell. *Alcaligenes, Acinetobacter, Aeromonas, Bacillus, Corynebacterium, Flavobacterium, Lactococcus, Lactobacillus, Pseudomonas* and *S. putrefaciens, Sarcina* are associated with proteolysis in fish.

The lipolytic microbes secrete lipolytic enzymes to the surrounding and hydrolyse fats into fatty acids and glycerol. Subsequently, they are utilised by microbes. The accumulation of free fatty acids in fish tissue leads to the development of rancid flavour in fish. This rancidity is called "hydrolytic rancidity". However, when compared to protein hydrolysis, lipid hydrolysis is negligible. Many of the proteolytic bacteria are found to be lipoplytic. Organisms of genera *Pseudomonas, Alcaligenes, Staphylococcus, Serratia* and *Micrococcus* are lipolytic bacteria.

3.2.4. Breakdown of Nucleotides

Hypoxanthine which cause bitter off flavour in fish is formed by autolytic or microbial activity. Several spoilage bacteria such as *Pseudomonas, Shewanella putrefaciens* and *Photobacterium phosphorium* produce hypoxanthine from inosine or inosine monophosphate.

Crustaceans

Microbial spoilage of shrimp is more prevalent than lobsters and crabs as the shrimps die immediately after harvest whereas crabs and lobsters remain alive until they are processed. Shrimps are reported to contain higher content of NPN compounds (free amino acid esp. arginine, TMAO) than fish and contains catheptic like enzymes that rapidly breakdown proteins. Initial spoilage of shrimp is accompanied by the production of large amounts of volatile base nitrogen formed from NPN extractives. Indole is formed from the amino acid tryptophan and is used as an indicator of shrimp spoilage. Many of the fish spoilage organisms are reported to cause spoilage in crustaceans with *Pseudomonas, Acinetobacter - Moraxella, Proteus* and *yeast* spp. being predominant on microbially spoiled crustacean meat. Crabs, shrimps and lobsters are also cooked prior to chilling/freezing to extend the shelf life. However spoilage occurs when they are exposed to contaminant that causes spoilage after cooking.

Table 3.5: Substrate and Off Odour/Off Flavour Compounds Produced by Bacteria during Spoilage of Fish

Substrate	*Degradation Products*
TMAO	TMA
Cysteine	H_2S
Methionine	CH_3SH, $(CH_3)_2S$
Carbohydrates and lactate	Acetate, CO_2, H_2O
Inosine, IMP	Hypoxanthine
Amino- acids (glycine, serine, leucine)	Esters, ketones, aldehydes
Amino- acids, urea	NH_3

Molluscs

The molluscan shellfish commonly consumed are oysters, clams and scallops. These animals differs from teleosts and crustaceans in having a significant content of carbohydrates (clams, scallops- 3.4 per cent and oysters-5.6 per cent) and a lower quantity of nitrogen in their flesh. The carbohydrate is largely in the form of glycogen, and so fermentative activities occur as part of microbial spoilage. Glycogen is metabolised by Lactic acid bacteria (*Lactobacillus* spp.), enterococci, and coliforms thus lowering the pH. Breakdown of the nitrogenous compounds by *Vibrio* and *Pseudomonas* results in the production of ammonia, amines and volatile fatty acids. The microbes of molluscs vary with the quality of water from which they are harvested and other post harvest actors. The following genera of bacteria have been recovered from spoiled oysters: *Serratia, Pseudomonas, Proteus, Clostridium, Bacillus, Escherichia, Enterobacter, Pseudoalteromonas, Shewanella, Lactobacillus, Flavobacterium* and *Micrococcus*. As spoilage sets in and progresses, *Pseudomonas* and *Acinetobacter-Moraxella* spp. predominate, with enterococci, lactobacilli and yeasts dominating the later stages of spoilage.

Cephalopods

The shelf life of cephalopods - octopus, squid and cuttle fish held in ice is relatively shorter than most fin fish of about 6-9 days. The non protein nitrogen content of cephalopods is high and the spoilage is characterised by ammoniacal off odours due to rapid onset of ammonia production. The bacteria involved in spoilage of cephalopods include *Pseudomonas, Shewanella, Acinetobacter, Flavobacterium* and *Moraxella*.

3.3. Microbiology of Chilled Fish

Microbial food spoilage is an area of global concern, as it has been estimated that more than 25 per cent of all food including seafood produced is lost post-harvest owing to microbial activity. Hence, it is very important to understand the growth and activity of spoilage microorganisms associated with aquatic foods, as it shall help to extend the shelf life of fish and shellfish by assisting with lowering the temperature with ice that reduce the losses by microbial spoilage and protect the consumers from food-borne illnesses.

Holding the fish in low temperature close to freezing point of water affects the microflora of fish. The low temperature reduces microbial and enzymatic activity resulting in extension of shelf life of fish. Factors such as quality of fish, method and duration of chilling and efficiency of storage method influence the quality of chilled fish. Many microorganisms associated with fish survive low temperature and spoilage is mainly caused by psychrotrophs. Among the bacterial flora dominance of Gram negatives is observed over Gram positive bacteria. Chill storage brings about changes in composition of microflora, and mesophiles are gradually dominated by psychrophiles. The low temperature, however, helps to maintain fish in good condition only for a short time. The prolonged storage under chill condition leads to growth of psychrophilic microorganisms resulting in spoilage of fish.

Low temperature preservation of food is based on the principle of reducing the microbial activity by subjecting to low temperature condition. As all metabolic activities of microorganisms are catalyzed by enzymes, and enzyme reactions are dependent on temperature, the low temperature slows down enzyme activity, thereby brings about reduction in microbial activity. The reduced microbial activity prolongs shelf life of foods.

Shelf-life of Temperate and Tropical Water Fish

Activity of microorganisms is known to decrease by two fold with every 10°C decrease in temperature. Fresh fish from temperate waters generally have prolonged shelflife than tropical fishes. This is because of the low ambient temperature in temperate waters coupled with low body temperature of fish, being cold blooded, lowers both enzymatic and bacterial growth, thereby reduces deteriorative change and helps in extending the shelflife. In tropical waters, the high ambient temperature and high body temperature of harvested fish promotes deteriorative changes by enzymes and microorganisms and thus hastens the spoilage.

Methods of Low Temperature Presentation

Low temperature preservation is generally attained by employing three different temperature conditions. They are;

- ☆ **Chilling temperature:** keeping foods at 10-15°C (slightly above refrigerated temperature).

- ☆ **Refrigerator temperature:** Keeping foods at 0~7°C
- ☆ **Freezer temperature:** Storage of foods below -18°C.

Chill Storage of Fish

Chill storage is a process by which temperature of fish is reduced close to freezing point of water (0°C). This delays both biochemical and bacteriological processes, thus prolongs shelf life. Deteriorative changes are retarded as long as low temperature is maintained. This ensures preserving natural nutritional and functional properties of food. Chilling can be achieved by use of ice (crushed/ flake ice) and use of homogenous coolant (cold air or cold liquid), and refrigerated temperature. Use of cold liquid may be in the form of chilled freshwater for light chilling or refrigerated seawater/brine to attain temperature of 0-1°C.

Bacteria Associated with Low Temperature Storage

Several microorganisms are capable of surviving and growing at low temperature and cause spoilage. Bacteria capable of a growing at or below 7°C are widely distributed. Gram negative bacteria are more common than Gram positive. Psychrotrophs grow well at this temperature. Growth at temperature below 0°C is caused mainly by yeasts and molds than bacteria because of low water activity. The lowest recorded temperature for growth of microorganisms in food is -34°C, by yeast.

The composition of microorganisms associated with fish changes during chill storage. Proportion of mesophiles decrease and psychrophiles dominate. Common bacterial genera associated with chilling temperature condition of foods are; *Acinetobacter, Aeromonas, Enterococcus, Pseudomonas, Vibrio, Erwinia, Moraxella, Enterobacter, Achromobacter, Flavobacterium, Micrococcus etc.* Extension of shelf life varying from 6 days to 30 days for different fish species has been reported by ice storage.

Table 3.6: Spoilage Activity of Bacteria

Spoilage Activity	*Microorganisms*
High	*Pseudomonas, Shewanella putrifaciens, Pseudomonas (Alteromonas) fluorescens, Fluorescent pseudomonads*
Moderate	*Moraxella, Acinetobacter and Alcaligenes*
Low	*Aerobacter, Lactobacillus, Flavobacterium, Micrococcus, Bacillus and Staphylococcus*

Fish use trimethylamine oxide (TMAO) as an osmoregulant to avoid dehydration in marine environments and tissue water logging in fresh water. Bacteria such as *Shewanella putrifaciens, Aeromonas* spp., Psychrotolerant Enterobacteriacceae, *Photobacterium phosphoreum* and *Vibrio* spp. can obtain energy by reducing TMAO to TMA creating the ammonia-like offflavors. TMA is used universally to determine microbial deterioration leading to fish spoilage. Lists spoilage bacteria in descending order of spoilage activity.

Bacteria on fish caught in temperate waters will enter the exponential growth phase almost immediately after the death of fish. This is also true when the fish are iced, probably because the microflora is already adapted to the chill temperatures. During ice storage, the bacteria will grow with a doubling time of approximately 1 day and, after 2-3 weeks, will reach numbers of 10^8-10^9 cfu/g flesh or cm^2 skin. During ambient storage, a slightly lower level of 10^7-10^8 cfu/g is reached in 24 h. The bacteria on fish caught in tropical waters will often pass through a log-phase of 1-2 weeks, if the fish are stored in ice, thereafter exponential growth begins. At spoilage, the bacterial level on tropical fish is similar to the levels found on temperate fish species. If iced fish are stored under anaerobic conditions or modified atmospheric storage, the number of the normal psychrotrophic bacteria such as *S. putrefaciens* and *Pseudomonas* are often much lower, *i.e.*, 10^6-10^7 cfu/g than on the aerobically stored fish. However, the level of psychrophilic bacteria such as *P. phosphoreum* reaches a level of 10^7-10^8 cfu/g when the fish spoil.

The composition of the microflora also changes quite dramatically during storage. Thus, under aerobic iced storage, the flora is composed almost exclusively of *Pseudomonas* spp. and *S. putrefaciens* after 1-2 weeks. At ambient temperature (25°C), the microflora at the point of spoilage is dominated by mesophilic Vibrionaceae and, particularly if the fish are caught in polluted waters, Enterobacteriaceae. A clear distinction should be made between the terms "spoilage flora" and "spoilage bacteria", since the first describes merely the bacteria present on the fish when it spoils whereas the latter is the specific group that produce the off-odours and off-flavours associated with spoilage.

Modified-atmosphere Packaging

A natural atmosphere rich in oxygen (21 per cent) is responsible for oxidative processes and for all aerobic respiratory life. Low oxygen levels have been shown to substantially prolong the freshness and quality life of refrigerated seafood products. MAP extends the shelf life of most fishery products by inhibiting bacterial growth and autoxidation.

In MAP, the natural atmosphere is replaced with a controlled gas mixture (carbon dioxide, nitrogen, oxygen *etc.*). Carbon dioxide is the most important gas in MAP of fish because of its bacteriostatic and fungistatic properties. In the absence of oxygen, partial fermentation of sugars occur leading to lower pH. Both carbon dioxide and low pH inhibit the growth of the typical spoilage bacteria such as *Pseudomonas* and *Shewanella*. Bacterial composition under MAP shifts from mostly Gram-negative to predominantly Gram-positive (lactic) bacteria. Brochothrix thermosphacta and psychrotrophic lactic acid bacteria (LAB) can produce spoilage characteristics; however, they are usually process contaminants, not part of the normal flora of the meat animals.

The principal effect of raised carbon dioxide-MAP is an extension of the 'lag' phase of the growth of the bacteria on the fish, the inhibition of common 'spoilage' bacteria (*Pseudomonas, Flavobacterium, Micrococcus* and *Moraxella*), and the promotion of a predominantly Gram-positive, slower-growing flora.

The single most important concern with respect to the use of MAP is the potential for outgrowth and toxin production by *C. botulinum*. Of particular concern are the psychrotrophic type E and non-proteolytic type B and F strains, as they are able to grow at temperatures as low as 3.3°C and produce toxins, without overt signs of spoilage. Growth and toxin production have been detected in artificially contaminated packs of whole trout after 1 week's incubation at 10°C.

3.4 Microbiology of Frozen Fish

Preservation by Freezing

Freezing involves lowering of temperature of food to -20°C and storage at same temperature. At this temperature the water in food as well as in microorganism is converted to ice crystals which affect fluidity of cell. This ensures prolonged shelflife as microbial activity is completely stopped at this temperature condition.

Freezing is Achieved by

- ☆ Quick freezing: where temperature is lowered to -20°C within 30 min.
- ☆ Slow freezing: where temperature is lowered to -20°C within 3~72 hours.

Quick freezing is more advantageous than slow freezing in achieving product quality. During freezing water in food is converted to ice crystals of variable size. Freezing also brings about changes in properties of food such as pH, titratable acidity, ionic strength, viscosity, osmotic pressure, freezing point, O/R potential *etc.* These changes along with non-availability of water make the environment unsuitable for microbial growth and activity.

Table 3.7: Comparison of Effect of Freezing Methods on Microorganisms

Quick Freezing	*Slow Freezing*
Small ice crystals formed	Large ice crystals formed
Suppresses microbial metabolism	Break down of metabolic rapport and causes cell damage
Brief exposure to adverse conditions	Longer exposure to injurious factors
No adaptation to low temperature	Gradual adaptation
Causes thermal shock to microbes	No thermal shock effect
No protective effect	Accumulation of concentrated solutes with beneficial effects.
Drip loss is less	Drip loss is more

Microbiology and Spoilage of Frozen Fish

The bacteriological quality of frozen products depends on the bacterial load of the raw material, contamination during handling and processing and extent of removal of these contaminants during processing. The freezing and storage under frozen condition has detrimental effect on surviving microorganisms and reduction in count is highly variable.

Growth and Survival of Microorganisms Associated with Frozen Foods

The psychrotrophic bacteria in fish are sensitive to freezing stress, and the sensitivity is strain dependent. Spoilage microorganisms generally grow and cause spoilage when the raw fish product is held long time before freezing, frozen at a very slow rate (slow freezing conditions), thawed too slowly or held under thawed condition for a long time. However, the ultimate activity of microorganisms depends on the time duration and temperature of holding the product. As most microorganisms are unable to grow below -10°C or -12°C, increase in temperature above this limit results in dramatic increase in growth rate. The changes in microflora and biochemical alterations observed in frozen fish products are similar to that of raw chilled fish. Seafood held at elevated frozen storage temperatures (-10°C to -5°C) are likely to support mold growth though at very slow rate. Some molds and yeasts can grow in that range. Though bacteria do not grow in frozen foods, they are able to survive freezing and frozen storage to a certain extent, so that thawed fish spoils about as fast as fish that has never been frozen. Gram negative bacteria are more readily killed by the freezing conditions than Gram positive bacteria while, spores are least sensitive and thus show better survival. Microbial analysis of frozen product gives some information about the quality of the fish before it was frozen. However, all strains of microorganisms are to some extent killed by freezing and frozen storage. Hence, the bacterial load in frozen fish is always lower than that observed before the product was frozen. For frozen foods use of *E. coli* as indicator of sanitary quality during processing is not suitable, and instead fecal streptococci are preferred owing to their greater ability to survive freezing.

Pathogenic Microorganisms and Frozen Foods

Among human pathogenic microorganisms, *Vibrio parahaemolyticus* is encountered in small numbers in frozen foods since it is quite sensitive to conditions of freezing and thawing. However, these potentially dangerous bacteria get destroyed on cooking, but pose threat only under the conditions of recontamination of the cooked food and abuse of holding time and temperature.

The potentially toxigenic *Clostridium botulinum* is not affected by freezing conditions and presents no hazard unless conditions for its outgrowth and toxin production are provided. However, toxins that might be present in the raw product or produced as a result of bacterial growth in seafood prior to freezing would not be inactivated by the freezing process. Generally, conditions permitting the development of large populations of spoilage bacteria also favour toxigenesis of

C. botulinum. The poor temperature control which is commonly encountered during the distribution of frozen seafood favours bacterial growth leading to spoilage. Staphylococci are reasonably resistant to the effects of freezing hence their load in frozen food gives an indication on the extent of handling/human contact that the food had received prioer to freezing.

Prevention of Spoilage

Foods frozen and held at appropriate temperature conditions are free from microbial spoilage unless the conditions of freezing and thawing had permitted growth of associated microorganisms as in slow freezing and slow thawing. Therefore, temperature control is the principal means to stop microbial activity in frozen seafood involving rapid freezing, holding at or below -18°C and rapid thawing

Shelf Life of Frozen Foods

Frozen foods can be stored indefinitely without microbial spoilage, but not done so as they lose original flavour and texture after thawing. Thus frozen foods are assigned freezer life. Freezer life for frozen foods is determined based on texture, flavor, tenderness, colour and nutritional quality upon thawing and cooking. Freezer life of frozen stored food does not depend on microbiology of frozen foods.

Effect of Freezing on Microorganisms

Freezing causes sudden mortality immediately on freezing, varying with microbial species. The number of surviving microorganisms after freezing die gradually when stored in frozen condition. Decline in microbial number is rapid at temperature below freezing point (-2°C) than at lower temperature, and is slow below -20°C. Bacteria differ greatly in their capacity to survive during freezing. Generally cocci are more resistant than Gram negative bacteria. Food poisoning bacteria are less resistant to freezing while, microbial endospores and toxins are not affected by freezing.

3.5 Microbiology of Cured Fish

Dried salted fish is consumed in many countries, especially in developing countries where they constitute an important source of low cost dietary protein. The preservation activity of drying is dependent on the reduction of water activity to stop or slow down the growth of spoilage microorganisms, as well as the occurrence and rate of chemical and enzymatic reactions. Fish and seafood are prone to rapid microbial spoilage; thus, adequate care must be taken in drying fish. The temperature and the treatment received before drying are important factors to produce safe products. Drying of highly perishable products, such as fish at low temperature, raises the risk of microorganisms and may produce an unsafe product. Salting is a preliminary step in the drying process and is critical to obtain a commercial product with an adequate shelf-life and good quality. Using salt before drying will enhance the capability of drying to reduce water activity to lower values within a short time. Osmotic dehydration is an important technology that enables both the removal of water from the product and the modification of its functional properties by impregnation of desired solutes. Osmotic dehydration is commonly performed by immersion of the product in salt solutions. It creates counter-current mass transfer fluxes of water and solutes, namely water outflow from the product to the surrounding solution and solute infusion into the product. In addition, leakage of the product solutes has an impact on the organoleptic properties and nutritional value of the product. The preservative effect of salting is mainly attributed to the decrease in water activity (a_w) inhibiting the growth of many spoilage microorganisms. In addition, chloride ions are toxic for some microorganisms.

Microbiology and Spoilage of Cured Fishery Products

Bacterial counts of fully dried seafood are generally low, unless there has been extensive surface contamination. Spoilage of such products is mainly caused by halophilic bacteria which can persist on the surfaces of contaminated cured seafood. However, the microbiological hazards due to pathogenic microorganisms are negligible in cured or salted seafood products. Smoked seafood products vary widely in microbial stability depending on the nature and degree of severity of processing. Heavily salted, hot smoked products are microbiologically similar to the fully dried products since their water content is too low to support bacterial growth and hence pose little or no hazard. Lightly smoked products that are brined only enough to improve a flavour carry a mixed microbial population and are only slightly more stable than unprocessed fish. Generally, Gram positive bacteria dominate the microflora of such products soon after preparation but Gram negative bacteria gradually become more numerous during refrigerated storage and are ultimately responsible for their spoilage. The hot smoked fishery products that have not been adequately dryed are of high risk due to selective outgrowth of *C. botulinum* resulting from favourable Eh and elimination of competing bacteria.

Categories of Dried Foods

Dried foods are generally divided in to 2 categories based on moisture content. They are, Low moisture foods: These foods do not contain moisture level more than 25 per cent, and have a_w of 0.6 or less. These foods have better keeping quality and shelflife.

Intermediate moisture foods (IMF): These contain 15-50 per cent moisture and aw between 0.60 and 0.85.Ex: Dried fruits, cakes *etc.*

Growth of most organisms is prevented at aw <0.65, but some molds (Ex: *Aspergillus echinulatus, Zygosaccharomyces rouxii*) are able to grow and cause spoilage. Some of the molds involved in the spoilage of dried foods are *Candida, Botrytis, Rhizopus, Mucor, Saccharomycopsis, Alternaria, Aspergillus, Zygosaccharomyces etc.* Though drying destroys some microorganisms, endospores of bacteria, yeasts, molds and certain bacteria survive drying process. Microbial spoilage of dried foods can be prevented by storing at low Rh condition. Storage at high Rh condition enables absorption of moisture from the atmosphere by dried foods until equilibrium is reached leading to spoilage beginning from the product surface.

Microorganisms Associated with Dried Fish

Spoilage of dried fish during storage can result from fungi and bacteria.Fungi are the common cause of spoilage of salt dried fish. Being aerobic and requiring low aw these grow on surface. Several species such as *Aspergillus, Penicillium, Fusarium, Paecilomyces etc.* have been recorded from dried fish in India.

Xerophilic Microorganism

Xerophilic mould is an extremophilic organism that can grow and reproduce in conditions with a low availability of water, also known as water activity. Water activity (aw) is a measure of the amount of water within a substrate an organism can use to support sexual growth. Xerophiles are often said to be "xerotolerant", meaning tolerant of dry conditions. They can survive in environments with water activity below 0.8.

As a group, xerophiles are extremely important in the spoilage of many processed foods and stored commodities, and in indoor environments. Moderate xerophiles include species within *Aspergillus, Penicillium* and *Eurotium*. Extreme xerophiles compete poorly at high aw, because they require decreased aw for growth. Some xerophiles have a preference for salt or sugar substrates, whereas other species can be isolated from both jam and salterns. Xerophiles are widely spread on the fungal tree of life.

Xerophiles in Dried Fish Products

Xerophilic mould that is capable of growing at reduced water activity, can contaminate salted fish from different sources such as the raw fish itself and the additives specially the salt. Fungus usually grows well on unsalted and salted dried fish, which has high moisture, content. Moulds usually grow at relative humidity above 75 per cent. The optimum temperature for growth is 30-35°C.

Aspergillus species is another most predominant mould in the environment and can affect most kinds of food; it is capable of producing mycotoxins which present a great public health hazard associated with salted fish. The most common species are Aspergillus species and A. glaucus species, which cause an objectionable flavour and textural changes in fish. The metabolism causes the release of moisture and increase a_w around the affected parts. It rapidly spread over at the surface and spoils fish depending on moisture level. Aflatoxins are produced in nature by Aspergillus flavus and Aspergillus parasiticus, the four major naturally produced aflatoxins are known as aflatoxin B1, B2, G1 and G2. Aflatoxins are both acutely and chronically toxic to animals, birds and Man, where they produce four distinct effects; acute liver damage, liver cirrhosis, induction of tumors and teratogenic effects. Xerophilic mould strains isolated from salted fish had the ability to produce mycotoxins, the most mycotoxin detected was stregmatocystin followed by B2 and G1.

Eurotiumams telodami, E. repens and *E. rubrum, Cladosporium, Trichoderma, Neurospora, Nigrospora, Mucor* are also common isolates from dried fish, together with a range of *Aspergilli, including A. flavus, A. niger, Aspergillus sydowii, Aspergillus wentii*, and *Aspergillus penicilloides*. Several *Penicillium* spp. were also isolated, most notably *Penicillium citrinum* and *Penicillium thomii. B. halophilica* was isolated infrequently at low levels; it was, however, able to grow on dried salted fish, with lower a_w limit for growth of 0.747, around the water activity of saturated sodium chloride.

The commonly occurring fungi in the west coast of India are *Aspergillus, Fusarium, Rhizopus* and *Mucor*. In cochin market, *A. flavus* and *A. ochraceus* are dominant where as in Tuticorin south east coast of India *Aspergillus* and *Penicilium* are dominant.

Prevention

Fish quality and preservative technique play an important role in the mycological quality of salted fish. Different means have been used to control fungal contamination and their formation of toxins. Organic acids and their salts are known to be efficient against microorganisms, particularly moulds. These substances are generally used in conservation of food materials without leaving residues that may cause health hazards to the consumers. Propionic acid and propionate formulations such as calcium and sodium salts are highly effective mould inhibitors commonly used in the food industry especially in cakes, bakery products and cheese. They have been listed as preservatives which generally recognized as safe food additives (GRAS). They are permitted in foods primarily as mould inhibitors, they tend to be highly specific against mould with the inhibitory action is being primarily fungistatic rather than fungicidal. Spoilage by mould was delayed and overall shelf life is improved by dipping of salted fish in propionic acid.

Bacterial Spoilage

Halophilic Microorganism

In salt dried fish (marine) bacteria are capable of growing at varying concentration of salt and can cause spoilage. These include;

- ☆ Slightly halophilic bacteria: These grow at 2~5 per cent salt. Ex: Many marine bacteria.
- ☆ Moderately halophilic bacteria: These grow at 5-20 per cent salt content. Ex: Many bacteria such as *Pseudomonas, Vibrio, Moraxella, Micrococcus, Acinetobacter, Flavobacterium etc.*
- ☆ Exstermely halophilic bacteria: These tolerate 20-32 per cent salt. Ex: *Halobacterium, Micrococcus, Sarcina etc.* Exstremely halophilic bacteria cause intense red/pink pigmentation in dried food.

- ✰ Halophic bacteria are capable of growing at low RH (75 per cent) unlike other bacteria. The main source of these bacteria to fish is through the salt used for salting of fish. These grow on surface of salt dried fish resulting in red coloration/pigmentation.

Halophilic Mould

In salted fish, brownish black or yellow brown spots are seen on the fleshy parts. This is mainly caused by growth of halophilic mould called Sporendonemaepizoum (*Sporendonemasebi, Wallamiasebi*). This gives the fish a very bad appearance. The growth depends upon the hygienic condition of the curing yard and storage premises. These moulds are harmless, do not damage the flesh and growth is very slow. It grows only, if fish absorbs moisture from the atmosphere. It can be prevented by good hygienic method in and around the processing plant. The presence of mould on the surface of the fish makes the product unacceptable to the consumer besides having the risk of toxin produced by some type of moulds on fish. The fish may be re-dried and red or damp the fish to prevent the contamination. During the initial stages of appearance of moulds on the fish, it is possible to remove them manually. In advanced stages when it has penetrated the flesh nothing can be done. To avoid this mould growth it is necessary that the fish be dried properly to pack the fish in required type of packaging material and keep it in a cool and dry place from moisture. Chemical method of prevention includes dipping the fish in a 5 per cent solution of Calcium propionate in saturated brine for 3-5 minutes depending upon the size of the fish. *Basipetospora halophila* (syn. *Oospora halophila*) and Polypaecilum pisce are two white halophilic moulds that are associated with spoilage of salted fish. These moulds sometimes covering the whole of the fish with a whitish, powdery growth.

Prevention

During the initial stages of appearance of moulds on the fish, it is possible to remove them manually. In advanced stages when it has penetrated the flesh nothing can be done. To avoid the mould growth it is necessary that the fish be dried properly to pack the fish in required type of packaging material and keep it in a cool and dry place from moisture. Chemical method of prevention includes dipping the fish in a 5 per cent solution of Calcium propionate in saturated brine for 3-5 minutes depending upon the size of the fish.

Halophilic Bacteria

Halophilic bacteria require a NaCl concentration of 20-25 g/100 g dried salt-cured fish product for growth and are known to proliferate in the temperature range of 10-50°C with optima around 37°C. A temperature-dependent growth of extremely halophilic microorganisms may occur during prolonged storage, particularly if the salt-cured fish is insufficiently dried. The source of such bacteria is salt. It is commonly found in tropical countries like India. They are aerobic and proteolytic in nature, grows best at 36°C by decomposing protein and giving out an ammoniacal odour. Halophilic microorganisms are often strongly proteolytic and produce off-odours and off-flavours. Both the endogenous proteolytic enzyme activities and the proteases originating from the halophilic microorganisms may result in increased levels of peptides and free amino acids in the muscle. These degradation products will function as substrates for further growth of halophilic microorganisms. Such growth limits the shelf life of the product by the formation of pink or red discolouration on surface of the cured or salted product that adversely affects the appearance. 'Pink' describes a condition associated with surface growth of the pigmented halophilic bacteria *Halococcus* spp. and *Halobacterium* spp. The bacteria form coloured colonies on the surface of salted fish prior to drying, or after drying if the fish is stored in excessive humidity. The metabolism of the organisms will produce off-odours, including hydrogen sulphide. The bacteria are osmosensitive and will lyse in water. At the pre-drying stage, it is possible to wash off the infected parts, if not too contaminated, and still produce a reasonable dried product. Bacteria from these two genera also grow sufficiently in brine and even salt to cause pink discolouration.

Sarcina littoralis, Pseudomonas salinaria are the important species involved in pink discoloration. Growth of these microorganisms may be observed as red spots or areas on the cut side of the dried salt-cured fish. Eventhough these bacteria are not harmful to human, such growth results in spoilage appears on the surface as slimy pink patches andred discolouration which limits the shelf life of the product. These organisms survive but not grow in salt water. They have a strong proteolytic action and latter cause indole and hydrogen sulphide and they require 25 to 30 per cent salt. On fish they very rapidly react and soften the flesh and has putrid smell and flavour and the fish become unfit for consumption. These red halophiles grow on moist surface.

The halophilic lactic acid microorganisms Tetragenococcus muriaticus and Tetragenococcus halophilus form histamine in salt-cured and fermented fish products. Halophilic histamine forming microorganisms were detected in fish products containing up to 15 g NaCl/100 g product. Accumulation of histamine in salted fish products may be affected by their free histidine content and distribution of halophilic histamine producing flora. Initial level of histidine in fish is low and the growth of the halophiles may increase this concentration and subsequently the possibility of histamine formation.

Prevention

According to Codex Alimentarius, the red discolouration is not accepted and the product is considered commercially unsuitable. To minimize growth of halophilic microorganisms, storage at low temperature is required, preferably below 8°C. The best way of controlling the organisms is to refrigerate the fish, as little if any growth will occur below 5°C on salted fish. Usage of good quality salt will avoid this. This spoilage is mostly found in heavily salted fish and absent in unsalted fish.

Smoking

Smoking today, as referred to above, is mainly a process for flavouring fish rather than a preservation technique. The production of smoke by wood or sawdust burning can be done in a mechanical kiln or in traditional chimney kilns. In the mechanical kiln, temperature and humidity can be automatically controlled and the process is faster because the directed air movement draws the smoke through the fish. It can be used for cold or hot smoking. The quality of the smoke is related to the type of wood used. Aroma and flavour are a blend of smoke components. Hardwoods are considered better because of the milder flavour they impart to the food. However, softwood colours the product quickly. Blends of various woods can be formulated to produce sawdust with specific flavours. It has been reported that the smoke from mahogany, white mangrove and abura has an inhibitory effect on some microorganisms, such as *Escherichia coli*, *Staph. aureus* and *Saccharomyces cerevisiae*.

Preventing the occurrence of carcinogens during smoking has been an issue for the past several years. Maintaining the temperatures of pyrolysis between 425 and 200°C, using electrostatic filtration of the smoke, and the use of smoke generated by superheated steam or liquid smoke, are some of the ways of reducing the polycyclic aromatic hydrocarbon (PAH) compounds.

All the smoked products, whether cold- or hot-smoked, after being smoked need to be chilled immediately to 4°C before being packed, or quick-frozen and stored at -18°C (2). Depending on the type of product, they can be vacuum packaged, shrink packaged, canned or bottled, or packed in retortable foil pouches, wooden boxes or master cartons.

Cold-Smoking

The major difference between cold- and hot-smoking is the temperature of the process. In cold-smoking, temperatures must be kept under 30°C throughout the process. This difference is crucial for safety issues because the flesh is not cooked. Thus, all cold-smoked products should be cooked before consumption.

The maintenance of temperatures below 30°C avoids protein coagulation, and the cold smoke is denser, thus allowing the deposit of smoke on the fish surface, giving the product the desired appearance and flavour. Drying should be minimal at the start, but it will increase towards the end of the process. The length of the smoking process for such products is much greater than for the hot-smoked fish, but a 'safety' (*i.e.* pasteurisation) temperature is not achieved in any stage of the process. Thus, temperatures and times used in processing cold-smoked fish are very favourable for the proliferation of food-spoilage and food-poisoning microorganisms.

During cold-smoking, there is no point in the process that can fully assure the absence of *L. monocytogenes*. Neither the smoking temperature nor the salt content is enough to kill the organism. Furthermore, refrigerated storage and vacuum packaging still allow its growth. Although other human pathogens that can grow at refrigeration temperature, such as *Yersinia enterocolitica*, can be isolated from seafood, they are more sensitive to sodium chloride and unlikely to grow in smoked products. If the refrigeration chain is maintained, risk from the presence of *Bacillus cereus* and *Clostridium perfringens* can also be excluded as potential hazards because of the relatively high temperatures they require to grow to, in order to produce toxin, and their sensitivity to salt. Pathogenic vibrios, typically mesophilic, are of concern when smoking fish from tropical waters.

To avoid growth of non-proteolytic *C. botulinum* type E the product must have at least 3 per cent salt in the water phase and be stored below 5°C, but the proteolytic types A and B, if present, can grow with salt levels of 10 per cent in the water phase, which underlines the danger of storage at abuse temperatures of >10°C.

Hot-smoking

Hot-smoking must use temperatures between 70 and 80°C at any stage of the process to ensure protein coagulation throughout the product. The protein must be set or denaturated with lower temperatures before the application of higher temperatures to avoid case-hardening. A typical temperature set is as follows: firstly, a drying period at 30°C; secondly, a partial cooking period at 50°C; thirdly, a final cooking period at 80°C. Length of the smoking process at each stage will depend on the species and the type of product.

There are several species that are usually hot-smoked and the process can vary markedly. Boiled or steamed clams, oysters and mussels are usually wet-salted in 50° brine for 5 minutes and hot-smoked at 80°C for less than 1 hour. Hot-smoked products made from white fish will usually keep better than those from fatty fish. Vacuum packaging retards the onset of rancidity when fatty fish is used, but seems, apart from the improved appearance, to have no further advantage over other packing methods for other fish.

The product is often allowed to go mouldy by incubating it in a chamber containing a natural fungal flora, including *Eurotium repens* and *Eurotium rubrum*, and many closely related Aspergilli (*Aspergillus ochraceus, Aspergillus oryzae, Aspergillus ostianus, Aspergillus tamarii*) and several *Penicillium* spp. (*Penicillum cyclopium* and *Penicillium puterellii*). Growth of these moulds on the tuna results in the extraction of moisture, break down of fats and the moderation of smoke flavour, resulting in a characteristic flavour. The product is removed from the drying cycle, rewetted and remoulded to give the desired shape and consistency.

Liquid Smoke

Liquid smoke in combination with sodium chloride in a hot-process smoke flavoured fish was found to be effective in preventing growth and toxin production associated with the growth of *C. botulinum* types A and E spores in several fish species that were stored at 25°C for 7 and 14 days. When liquid smoke was used, the salt level could effectively be reduced from 4.6 to 2.8 per cent and still provide protection for 7 days. The synergistic effect of liquid smoke with other preservatives could reduce the amount of salt in the product when light diets are necessary whilst still meeting the necessary safety criteria. Liquid smoke was considered to provide advantages of safety, consistency of flavour and colour, nutritional neutrality and antioxidant activity. There are two groups of chemicals of concern in smoke: polycyclic aromatic hydrocarbons (PAHs) and nitrosamines, both of which are considered potential carcinogens. It has been shown that liquid smoke can reduce the formation of PAHs. The European Scientific Commitee for Food, established in its guidelines that smoke flavourings added to the food product must assure a level lower than 0.03 μg/kg of benzopyrene, and less than 0.06 μg/kg of benzoanthracene in the final foodstuff.

Pathogens: Growth and Survival

Clostridium botulinum

In the last few decades, popular demand for healthy foods has led to a change in the tastes preferred by the public and, with this change, pressure on the traditional methods of curing. Before, the products were heavily salted, smoked and dried, and *C. botulinum*, when present in the raw material, could not grow and produce toxin. A number of surveys have shown that fish from temperate waters is mainly contaminated with the psychrotrophic *C. botulinum* type E, but the presence of the mesophilic types A and B is of concern in fish from warmer waters. The presence of salt in cured products has a great effect on the growth of the bacteria, but the concentration of salt in smoked salmon or trout (usually varying from 1 - 4 per cent salt) is not high enough to prevent growth. The concentration required to prevent growth at room temperature can vary from as low as 3.5 per cent to 5 per cent, although approximately 10 per cent salt in the water phase is necessary to fully inhibit the mesophilic types A and B strains. It has been shown that, under optimal conditions, the non-proteolytic type E strains can grow in up to 5 per cent sodium chloride, but if the fish products are stored at temperatures <10°C, 3 per cent sodium chloride in the water phase is enough to inhibit growth of type E for at least 30 days. However, although the combination of the two factors, temperature δ5°C and salt ε3 per cent in the water phase is sufficient to prevent any growth of type E, types A and B are not affected by 3 - 5 per cent of salt in the water phase and great care must be taken to avoid any temperatures >10°C.

Sodium nitrite in conjunction with sodium chloride is known for its ability to inhibit *C. botulinum* type E and production of toxin. In the United States, sodiumnitrite can be added to the brine but this is not permitted in EU countries because nitrites might lead to formation of the carcinogenic nitrosamines. The FDA established that hot- or cold-smoked fish should have at least 3.5 per cent salt in the water phase, but, if it contains sodium nitrite ε100 ppm, it should contain not less than 3.0 per cent salt in the water phase.

Staphylococcus aureus

This is a ubiquitous organism but its main reservoir and habitat is animals and, more importantly, the human nose, throat and skin. Thus, if the person handling the product is a carrier of *Staph. aureus*, the organism may contaminate the product. The human carrier rate can be as high as 60 per cent of healthy individuals with an average of 25 - 30 per cent of the population positive for enterotoxin-producing strains. The organism is mesophilic with a minimum growth temperature of 7°C, but requires higher temperatures for toxin production (>10°C). It is a halotolerant bacterium able to grow at water activities as low as 0.86 and 10 per cent sodium chloride, and the minimum pH for growth is 4.0 - 4.5. In foodstuffs, these parameters change as a result of the interference of food components and other limiting factors that can act in an additive or synergistic way to prevent *Staph. aureus* growth. As they are poor competitors, they do not grow well in the rawmaterial, but, after processing, when the bacterial load of the product is lower, if recontamination occurs and storage conditions are favourable, they will grow rapidly. Sliced smoked fish products, as they require more handling, are particularly susceptible to contamination and, if storage temperatures are over 10°C, growth and toxin formation may occur. Once the product is contaminated, the organism can survive most of the light curing processes.

Enterobacteriaceae

The pathogens *Salmonella, Shigella* and *E. coli* all occur on fish products as a result of contamination from animal/human reservoirs. As with other organisms, the risk of infection can be eliminated by proper cooking. However, they can survive lightly cured processes when the sodium chloride content is less than 5 per cent and water activities are higher than 0.94. The degree of concern in cold-smoked salmon is low unless temperature abuse >10°C occurs.

Listeria monocytogenes

The psychrotrophic *L. monocytogenes* is a ubiquitous organism that has been detected in fresh fish skin, gills and guts as well as in several lightly preserved fishproducts such as cold-smoked fish, marinades (ceviche) and gravad fish. Its limiting conditions for growth are 0 - 45°C, salt levels of 8 - 12 per cent, pH 4.8 - 9.6 and water activities below 0.92 - 0.95.

Unlike some European countries, which considered *L. monocytogenes* pathogenic for only specific segments at risk, the USA established 'zero tolerance' for the organism, which means that the process must conform to a level of 'no detectable *L. monocytogenes*' in 25 g of the finished product.

Parasites

Despite the presence of parasites in fish being very common, most of them are of little concern to public health. They have complicated life cycles, which include intermediate hosts. As they are heat-sensitive, the only concern to public health is when eating raw or uncooked fish products. Thus, a number of fish products, such as 'lightly preserved' fish products, are considered unsafe. This includes matjes herring, gravad fish, cold-smoked fish, lightly salted caviar, and other local traditional products.

Of microbiological concern in the northern hemisphere are two parasites - the round worm *Anisakis* spp. from seawater fish, and the tape worm *Diphyllobothrium* spp. from freshwater fish, both being detected in cold-smokedsalmon.

3.6 Microbiology of Canned Fish

Microbiology and Spoilage of Canned Fishery Products

Canned seafoods are expected to be commercially sterile and free from spoilage and potentially pathogenic microorganisms. The bacteriological hazards of canned food mainly results from improper or inadequate processing or leakage of cans. The inadequate heat processing causes survival and growth of heat resistant clostridial spores responsible for botulism. The mesophilic spoilage organisms entering the cans through leakage caused due to improper seaming of cans grow during storage eventually resulting in spoilage of canned product.

The semi-processed canned seafood products are often subjected to bacteriological problems. The stability and safety of these products depend on the factors such as combination of preservatives used and pasteurization process applied. In most pickled products that depend on salt (*e.g.* anchovies) or a low pH (*e.g.* mussel) for stability, the heat process given destroys both hazardous and spoilage microorganisms. Yeast like Pichia fermentans often cause illness after growing in semi-processed canned seafood.

In some smoked products which are canned using minimal heat treatment, the storage stability is attributed to presence of salt, smoke constituents and a low water activity. Thus, production of safe final product can be achieved only by having good control over processing parameters.

Food Preservation by High Temperature

The high temperature preservation of food is based on destructive effect of heat on microorganisms, thereby extending shelf life of foods. High temperature refers to any temperature higher than ambient temperature applied to food. Preservation of foods by heat treatment can be done by three methods *viz.* pasteurization, sterilization and cooking.

Pasteurization

Pasteurization refers to use of heat at the range of 60~80°C for a few minutes for the elimination/destruction of all disease causing microorganisms, and reduction of potential spoilage organisms. Pasteurization is commonly used in the preservation of milk, fruit juices, pickles, sauces, beer *etc.*Pasteurization process which is commonly employed in milk preservation can be achieved by heating the milk at 63°C for 30 min, called low temperature long time (LTLT) process; or 72°C for 15 sec, called high temperature short time (HTST) process. This process destroys most heat resistant non-spore forming pathogens (Ex. *Mycobacterium tuberculosis*), all yeasts, molds, Gram negative bacteria and most Gram positive bacteria.

Organisms Surviving Pasteurization

The pasteurization treatment does not bring about complete removal or destruction of microorganisms and some organisms survive pasteurization process. The surviving organisms are of two types:

- ☆ Thermoduric microorganisms
- ☆ Thermophilic microorganisms.

Thermoduric microorganisms are those which can survive exposure to relativity high temperature but do not grow at these temperatures. Example: The non-spore forming *Streptococcus* and *Lactobacillus* sp. can grow and cause spoilage at normal temperature. So, milk need to be refrigerated after pasteurization to prevent spoilage. Thermophilic organisms are those that not only survive high temperature treatment but require high temperature for their growth and metabolic activities. Example: *Bacillus, Clostridium, Alicyclobacillus, Geobacillus etc.*

Sterilization

Sterilization or appertization refers to destruction of all viable organisms in food as measured by an appropriate enumeration method. This process kills all viable pathogenic and spoilage organisms. However, organisms that survive are non-pathogenic and unable to develop in product under normal conditions of storage. Thus, sterilized products have long shelf life. Commercially sterile or commercial sterility is often used for canned foods to indicate the absence of viable microorganisms detectable by culture methods or the number of survivors is so low that they are of no significance under condition of canning and storage.

Processing of food for preservation using high temperature depends on the physical nature of the food. Foods (solid or semisolid) are generally processed by packing in cans, sealing and then sterilized. Most liquid foods are sterilized, packed in suitable containers and sealed aseptically. Temperature and time of sterilization given to a food depends on the nature (pH, physical state, nutritional type *etc.*) of the food being processed.

Heat Resistance of Microorganisms

- ☆ Heat resistance of microorganisms is related to their optimal growth temperature.
- ☆ Psychrophiles are most sensitive and thermophiles are most resistant to heat treatment.
- ☆ Spore forming microorganisms are most resistant than non spare formers.
- ☆ Gram positive bacteria are more resistant than Gram negative bacteria.
- ☆ Yeasts and molds are fairly heat sensitive
- ☆ Spores of molds are slightly more heat resistant than vegetative cells.

Heat Resistance of Spores

Bacterial spores are more heat resistant than vegetative cells. Thermophiles produce more heat resistant spores. Since spore inactivation is the main concern in canned foods, high process temperature is used to achieve this. The heat resistance of bacterial endospore is due to their ability to maintain very low water content in the DNA containing protoplast. Presence of calcium and dipicolonic acid in high concentration in spores helps to reduce cytoplasmic water. Higher the degree of spore dehydration greater will be its heat resistance.

Factors Affecting Heat Destruction of Microorganisms

Several factors associated with microorganisms as well as their environment affect heat destruction of microorganisms.

Water

Heat resistance of microorganisms increases with decrease in moisture/water activity and humidity. This is due to faster denaturation of protein in presence of water than air.

Fat

Heat resistance increases in presence of fat due to direct effect of fat on cell moisture. Heat protective effect of long chain fatty acids is better than short chain fatty acids.

Salts

Effect of salt in heat resistance of microorganisms is variable, and depends on type of salt, concentration used, and other factors. Some salts (sodium salts) have protective effect on microorganisms and others (Ca2+ and Mg2+) make cells more sensitive. Some salts (Ca and Mg) increase water activity, while others (Na+) decrease water activity there by affecting heat sensitivity.

Carbohydrates

Presence of sugars in suspending medium increases heat resistance of microorganisms due to decreased a_w. Different sugars show varying effect. Heat resistance decreases in the order of; sucrose>glucose>sorbitol>fructose>glycerol.

pH

Microorganisms are most heat resistant to heat at their optimum pH for growth (about pH 7-0). Increase or decrease in pH reduces heat sensitivity. Thus, high acid foods require less heat processing than low acid foods.

Proteins

Proteins have protective effect on microorganisms. As a result high protein foods need a higher heat treatment than low protein foods to obtain similar results.

- ☆ **Number of microorganisms:** Larger the number of microorganisms, higher the degree of heat resistance. This is due to the production of protective substance excreted by bacterial cells, and natural variations in a microbial population to heat resistance.
- ☆ **Age of microorganisms:** Microorganisms are most resistant to heat in stationery growth and least in logarithmic growth phase. Also old bacterial spores are more resistant that young spores.
- ☆ **Growth temperature:** Heat resistant of microorganisms increases with increase in incubation temperature, especially in spore formers. This is mainly related to genetic selection favoring growth of heat resistant forms. Cultures grown at 44°C are known to have three times more heat resistance than those grown at 35°C.
- ☆ **Inhibiting compounds:** Heat resistance of most microorganisms decreases in the presence of heat resistant microbial inhibitor such as antibiotic (nisin), sulfur dioxide *etc.* Heat and inhibiting substances together are more effective in controlling spoilage of foods than either alone.
- ☆ **Time and temperature:** Generally believed that the longer the heating time, greater the killing effect. But higher the temperature, greater will be the killing effect. Thus, as temperature increases, time necessary to achieve the same effect decreases. Also, the size and composition of containers affect heat penetration.

Thermal Destruction of Microorganisms

The preservative effect of high temperature treatment depends on the extent of destruction of microorganisms. Certain basic concepts are associated with the thermal destruction of microorganisms.These include;

- ☆ Thermal death time (TDT)
- ☆ D- value
- ☆ Z- value
- ☆ F- value
- ☆ 12D concept

Thermal Death Time (TDT)

TDT is the time required to kill a given number of organisms at a specified temperature. Here, temperature is kept constant and the time necessary to kill all cells is determined. Whereas, thermal death point is the temperature necessary to kill given number of organisms in a fixed time, usually 10 min. But it is of less importance. TDT is determined by placing a known number of bacterial cells/ spores in sealed containers, heating in a oil bath for required time and cooling quickly. The number of survivors from each test period is determined by plating on a suitable growth media. Death is defined as the inability of organism to form viable colonies after incubation.

D-value (Decimal reduction time)

D-value is the time in minutes required at specified temperature to kill 90 per cent of microorganisms thereby reducing the count by 1 log units. Hence D - value is the measure of death rate of microorganisms. It reflects the resistance of an organism to a specific temperature and can be used to compare the relative heat resistance among different organisms/spores. D-value for the same organism varies depending on the food type. D -value is lower in acid foods and higher in presence of high proteins.

Example

- ☆ 250°F (121.1°C) for
  stearothermophilus: 4-5 min
  botulinum: 0.1 – 0.2 min.
- ☆ 95°C for
  coagulans: 13.7 min
  licheniformis: 5.1 min.

Z-Value

Z-value refers to degrees of Fahrheit required for the thermal destruction curve to drop by one log cycle. Z value gives information on the relative resistance of an organism for different destruction temperature. It helps to determine equivalent thermal process at different temperature.

Example

If adequate heat process is achieved at 150°F for 3 min and Z -value was determined as 100°F, which means the10°F rise in temperature reduces microorganisms by 1 log unit. Therefore, at 140°F, heat process need to be for 30 min and at 160°F for 0.3 min to ensure adequate process.

F-Value

F-value is the better way of expressing TDT. F- is the time in minutes required to kill all spores/vegetative cells at 250°F (121.0°C). It is the capacity of heat process to reduce the number of spores or vegetarian cells of an organism.

F-Value is calculated by:

$$Fo = Dr (\log a - \log b)$$

D = Decimal reduction time (D value)

= Initial cell numbers

= Final cell numbers

12D Concept

12D concept is used mainly in low acid canned foods (pH >4.6) where *C. botulinum* is a serious concern. 12D concept refers to thermal processing requirements designed to reduce the probability of survival of the most heat resistant *C. botulinum* spores to 10-

- This helps to determine the time required at process temperature of 121°C to reduce spores of *C. botulinum* to 1 spore in only 1of 1 billion containers (with an assumption that each container of food containing only 1 spore of *C. botulinum*).

I. Microbial Spoilage Associated with Canned Products

Typical microbial groups that have been isolated from canned fish and capable of causing spoilage include anaerobes, facultative organisms, spores, thermophilic organisms and micrococci, such as *Micrococcus*, *Streptococcus* and other Gram-positive cocci. The Bacillus group causes spoilage of canned fish products, not necessarily producing gas but characterised by off-odours or changes in the colour and texture of the fish flesh. In particular, *Bacillus cereus*, *Bacillus subtilis*, *Bacillus coagulans* and *Bacillus circulans* have been associated with products such as salmon, crab and shrimp. These may be more likely to dominate in minimally processed products.

The most common types of anaerobes that could cause gaseous spoilage in canned fish are the mesophilic clostridia. This spoilage is usually characterised by the visible disintegration of the flesh and subsequent off-odour. The gaseous spoilage can be minimized in products with reduced pH; however, the visible signs of spoilage usually persist. The optimum temperature for most mesophiles is 37°C; however, the range for growth can extend below 20 and above 50°C. The key proteolytic and often putrefactive clostridia relevant to fish spoilage are *Clostridium sporogenes* and *Clostridium bifermentans*. For those fish products that may be combined with carbohydrates, the saccharolytic group, include *Clostridium butyricum*, *Clostridium pasteurianum* and *Clostridium perfringens*. The most common cause of putrefactive spoilage in canned fish is *Clostridium sporogenes*.

1. Pre-process Spoilage Issues

There are many obvious spoilage issues related to the inadequate storage of raw materials pre-process. The two key issues that are specific to the manufacture of canned fish are histamine poisoning and staphylococcal food poisoning.

i. Histamine Poisoning

A specific form of food poisoning can be associated with certain types of canned scombroid fish. Histamine is produced in fish by the decarboxylation of histidine. This reaction is enhanced by the enzyme, histidine decarboxylase, which is present in many bacteria. Many fish species, particularly tuna and mackerel, contain large amounts of free histidine, which is then available for decarboxylation. Consequently, the combination of high levels of these bacteria because of poor hygienic practices and inadequate storage conditions can allow high levels of histidine decarboxylase activity and subsequent presence of histamine in canned fish. This can be minimised by the use of good-quality raw materials, hygienic handling and chilled storage of the fish prior to canning.

ii. Staphylococcal Food Poisoning

If *Staphylococcus aureus* was present in fish prior to processing and it was held at conditions suitable for growth, for example, at ambient/room temperature for greater than 3 hours, the organism could grow to sufficiently high concentrations to produce enterotoxins in the fish product. Growth and enterotoxin production can occur over the temperature range 10–45°C. The organism itself may not remain viable after processing as it is sensitive to heat and chemicals and indeed is unlikely to survive a typical $F_0$10 fish sterilisation process. The enterotoxins which are produced by this organism are, however, very heat stable and could remain throughout the heat processing and during shelf-life of the canned fish

2. Under Processing Issues

The reliability of the scheduled manufacturing heat process depends on the appropriate choice of hold time and temperature, particularly in relation to the expected effects on the target organism. All of the following actions are very important:

- The application of the heat to achieve the selected temperature and time combination;
- The control of the temperature for the required time; and
- The recording of these data to prove that the minimum process was achieved.

If there is a failure to achieve the scheduled heat process, there will be a risk of the survival of food poisoning and spoilage organisms, which are likely to be dominated by the presence of bacterial spores as opposed to vegetative bacteria. Spoilt swollen fish cans are usually associated with gas production from contaminating microorganisms, often anaerobic organisms, which have survived the heat process.

3. Post-process Spoilage Issues

Post-process spoilage accounts for 60–80 per cent of the spoilage of canned foods. Key factors are highlighted below that are relevant to and could be the cause of post-process microbial spoilage.

i. Poor Cooling after the Heat Process

When the cooling period is extensive; the final cooling temperature is not low enough; and the cans are transferred to a storage facility where the temperature is sufficiently high to allow the germination and growth of the spoilage organisms

ii. Leakage into the Cans

Spoilage microorganisms can enter through wet can seams, poor-quality seams with faults; through damaged containers with pinholes; issues with the side seam (*i.e.* tightess), over flanging and defective double seam or damage to the can during transport within the factory environment and can increase the risk of problems. When cans are crashed together or against another solid object it can damage the seams. If the can integrity is challenged, microorganisms could leak into the cans, grow in the product and cause issues. A wide range of microorganisms having potential to leak into the cans, include vegetative bacteria, spores, yeasts and moulds.

iii. Handling of Wet Containers

It is very important that cans are dried before any handling. Poor-quality cooling water and presence of bacteria in the cooling water can allow entry of microorganisms into the can. This can be avoided by addition of chlorine to a level of 2–5 ppm in the cooling water.

Ia. Thermal Process Requirements for Canned Products

In order to make canned food absolutely safe, thermal processes given must be sufficient to eliminate all pathogens. *Clostridium botulinum* type E is most prevalent in the freshly caught fish. *Clostridium botulinum* is able to reproduce inside the sealed container and can lead to the development of a potentially lethal toxin. *Clostridium botulinum* is hence taken as the target organism to device thermal processing requirements in canned foods. Bacteria, when subjected to moist heat at lethal temperatures undergo a logarithmic order of death. Survivor curve depicts the bacterial spores being killed by heat at constant lethal temperature.

To prevent microbial deterioration in the product, two factors that must be considered include the consumer safety from botulism, and the risk of non-pathogenic spoilage bacteria. A process equivalent to 12D reductions in the population of *C. botulinum* spores is sufficient for safety and this is referred to as a 12D process Such a low probability of survival is commercially acceptable, as it does not represent a significant health risk. Spoilage by non-pathogenic bacteria, will eventually threaten the profitability and commercial viability of a canning operation. It is important to quantify the maximum tolerable spore survival levels for the canned products. For mesophilic spores, other than those of *C. botulinum*, a 5D process is found adequate; while for thermophilic spores, process adequacy is generally assessed in terms of the probability of spore survival, which is judged commercially acceptable. It is generally found acceptable, if thermophilic spore levels are reduced to around 10^{-2} to 10^{-3}/g. There are two reasons for the tolerance of the higher risks of spoilage bacteria. First is at the storage temperatures of less than 35°C, the survivors will not germinate; and second is even if spoilage does arise, it will not endanger public health. If a thermal process is sufficient to fulfil the criteria of safety and prevention of non-pathogenic spoilage under normal conditions of transport and storage, the product is said to be "commercially sterile". There, the minimum thermal process required to provide safety from the survival of *C. botulinum* (*i.e.* 12D process) is equivalent to 2.8 min at 121.1°C at the slowest heating point (the SHP) of the container. This process is commonly referred to as a "botulinum cook".

Once the minimum process is established, a processing time and temperature regime have to be selected to reduce the numbers of spore forming microorganisms to an acceptable level. For instance, in order to prevent commercial losses through thermophilic spoilage by *C. thermosaccharolyticum* the thermal process must be equivalent, to 16 min at 121.1°C at the SHP of the container. The Fo value of a thermal process can be determined by two means: microbiological or physical. Microbiological method relies on quantifying the destructive effect of heating on bacterial numbers through their enumeration before and after thermal process; physical method measures the change in temperature during thermal process at the SHP of the container using thermocouples and relates this to the rate of thermal destruction at a reference temperature. The heat penetration data collected are treated in a number of ways in order to calculate the Fo value of the process.

II. Microbial Spoilage Associated with Cooked Products

Cooking reduces the microbial load of the products substantially. A number of biological hazards are relevant and important to consider, when the safety of cooked seafood needs to be evaluated. *L. monocytogenes* represents the main risk in some chilled and cooked seafood but typical product characteristics and storage conditions of cold-smoked and gravad fish are insufficient to prevent growth of this pathogen. Potential growth and toxin production by *C. botulinum* is another risk associated with seafood; however, it is possible to control this pathogen by environmental parameters *e.g.* chill storage and salt. Viruses, parasites, non-indigenous bacteria (*e.g. Salmonella*) and biogenic amines are the other risks.

Thermal Process Requirements for Cooked Products

One reason for cooking products is to eliminate vegetative cells of pathogenic bacteria that may have been introduced to the product by raw materials or by processing that occurs before the cooking step. Selection of the target pathogen is critical to find the effectiveness of cooking. Generally, *L. monocytogenes* is selected as the target pathogen because it is regarded as the most heat-tolerant, food borne bacterial pathogen that does not form spores. Cooking processes are not usually designed to eliminate the spores of bacterial pathogens and hence a 6D process is sufficient.

Cooking is sometimes performed on products immediately before placement in vacuum packaging or modified atmosphere packaging. These products include cooked, hot-filled soups, chowders, or sauces that are filled directly from the cook kettle using sanitary, automated, continuous filling systems designed to minimize risk of recontamination. They are often marketed under refrigeration, which is important for the control of *C. botulinum* type A and proteolytic types B and F. The cooking process for these products should be sufficient to eliminate the spores of *C. botulinum* type E and non-proteolytic types B and F. This is the case when the product does not contain other barriers that are sufficient to prevent growth and toxin formation by this pathogen. Generally, a 6D process is suitable. These products should be refrigerated or frozen to control *C. botulinum* type A and proteolytic types B and F. Control of refrigeration is critical to the safety of these products.

3.7 Microbiology of Fermented

Definitions

As applied to fish products, the term 'fermented' describes a spectrum of processes that ranges from the largely autolytic degradation of fish protein, to those processes in which the activity of lactic acid bacteria (LAB) plays an important part. The relative importance of these two activities depends on the formulation of the product. Those to which only salt is added tend to be mainly autolytic, producing the fish sauces and pastes, whereas in those where a carbohydrate source such as rice or sugar is also added, lactic fermentation can play a significant role.

Fermented fish products are largely confined to east and south-east Asia, though some are produced elsewhere. As with most traditional products that are still produced principally on a cottage industry or domestic scale, there are numerous variants of some common themes and a host of local names used to describe them. Essentially, they can be divided into two categories: fish/salt products and fish/salt/carbohydrate products.

Table 3.8: Fish Sauces and Pastes of South-East Asia

Country	*Sauce*	*Paste*
Burma	Amber/brown liquid; salty taste, cheese-like aroma	Red/brown salty paste
	Ngapi	nga-ngapi
Indonesia	ketjap-ikan	trassi-ikan
Kampuchea		trassi-udang (shrimps)
	nuoc-mam	prahoc (shrimps)
Korea	nuoc-mam-gau-ca (livers only)	mam-ruoc
		myulchljeot
Laos		jeots, jeotkals (shrimps)
		nam-pla (pa) padec
Malaysia	Budu	Belachan (shrimps)
Philippines	Patis	bagoong
Thailand	nam-pla	kapi
Vietnam	nuoc-mam	mam-ca (shrimps) man-tom

Fish/salt products such as the fish pastes and sauces tend to contain relatively high levels of salt, typically in the range 15 - 25 per cent and are used mainly as a condiment. The fish sauces are produced on the largest scale; they are exported to European, North American and other markets and are, in economic terms, the most important of all fermented fish products. The principal fish sauces and pastes of south-east Asia and their local names are listed in Tables 3.8 and 3.9

Table 3.9: Fish/Salt/Carbohydrate Products of South-East and East Asia

Country	*Products*
Japan	*I-sushi, e.g. ayu-sushi, funa-suchi, tai-suchi*
Kampuchea	Phaak, mam-chao, mam-seeing
Korea	Sikhae
Laos	Som-kay-pa-eun, som-pa, mam-pa-kor, pa-chao, pa-khem, som-pa-keng
Malaysia	Pekasam, cencalok
Philippines	*Burong-isda, e.g. burong-ayungi, burong-dalag, burong-bangus*
Thailand	Pla-ra, pla-som, pla-chao, som-fak/som-fug

Carbohydrates are added to some fish sauces during their preparation. For example, the best-known of the fish sauces (gyoshoyu) of Japan is shottsuru. It is produced from the sandfish Arctoscopus japonicus and has the rice-based enzyme preparation, koji, added during preparation. Sugar can also be added to the Indonesian fish sauce bakasang and the Thai product nam-budu.

Initial Microflora

The initial microbial levels are far from uniform, and counts ranging from below 104/g to in excess of 107 have been reported.Where other ingredients are used, they can contribute to the initial microflora. The mixed microflora of moulds and yeasts, principally species of the genera Rhizopus, Mucor, Hansenula, Endomycopsis and Saccharomyces - organisms that play an important part in the production process but can also contribute to the ultimate spoilage of the product.

Fish Sauces and Pastes

The amount of salt used is often only vaguely defined but is generally sufficient to reduce the a_w below levels at which most bacteria associated with the raw material will grow and, in many cases, sufficient to saturate the water phase present and produce an a_w of 0.74 or below. This, and the anaerobic conditions that prevail in the bulk of the material, mean that microorganisms have been generally held to play little or no role in the process that follows. The number of bacteria declines rapidly during the production of the Thai fish sauce nam pla and Patis comprising mainly *Halobacterium, Halococcus* spp. *Micrococcus* and *Bacillus* spp. The proteolytic activity during the first month of fermentation, suggesting that halophilic bacteria is play an important role in the production of fish sauce. Increasing the proportion of carbohydrate in the formulation decreases the product's buffering capacity and provides the LAB with more substrate on which to act. Rice is the most common carbohydrate source used. When added as boiled or roasted rice alone, efficient lactic acid production will depend upon the presence of amylolytic LAB or saccharification by other microorganisms or enzymes that may be present. Although the ability to ferment starch is not widespread in the LAB, a number of amylolytic species have been described and an amylolytic strain of *Lactobacillus plantarum* has been isolated.

Pathogens: Growth and Survival

Many fermented fish products are stored at ambient temperatures, typically 30°C, and consumed without cooking. Their safety therefore is critically dependent on the quality of the raw materials used and the inhibitory effects of the fermentation process.

There is little in the way of documented information associating fermented fish products with foodborne illness. In the fish sauces and pastes, the high salt concentrations preclude the growth of all food-poisoning bacteria and those initially present would not survive well the prolonged exposure to high salt during production. Improper storage of fish prior to processing could result in the production of toxins capable of persisting through to the product, but we are unaware of any reports of this happening.

The fish/salt/carbohydrate products depend on the combination of salt and pH/lactic acid for their safety. After capture, fish can potentially be contaminated with almost any foodborne pathogen as a result of improper handling and storage, and failure to achieve a sufficient combination of inhibitory factors could lead to their growth or survival. The bacterial pathogens most commonly associated with the living environment of fish, however, are *Vibrio parahaemolyticus* and *Clostridium botulinum*.

V. parahaemolyticus has been reported growing down to a pH of 4.8 and at salt levels as high 10 per cent, although these limits apply when all other factors are optimal. The combinations of pH and salt in many fermented products are likely to preclude growth of the organism completely and, under conditions inimical to growth and high ambient storage temperatures, viable numbers will probably decline quite rapidly.

Botulism has also been associated with a number of traditional foods. The more widespread current use of non-traditional containers such as glass or plastic jars is thought to be one contributing factor. Salting levels are often insufficiently low to prevent growth of *C. botulinum*, and warm weather can result in rapid growth of *C. botulinum* and the production of toxin, which will persist into the product, which is generally consumed raw.

Foodborne trematode infections are emerging as a major public health problem, affecting people, particularly in south-east Asia and the western Pacific regions. Infection with the trematode *Opisthorchis viverrini* is associated with cholangiocarcinoma. An investigation of survival of the fish parasitic nematode Anisakis in marinated herring has given some indication. With this particular organism, salt content was found to have the greatest impact on the viability of larvae, while increasing the acetic acid content of the formulation had little or no effect. One study with the trematode *Haplorchis taichui*, showed that inclusion of fermented fish in the Thai dish lab-pla improved inactivation of the metacercariae.

3.8. Microbiology of Irradiated Fish Product

Food Preservation by Use of Radiation

Radiations have potential application in food preservation because of their destructive effect on microorganisms. Radiation is defined as the emission and propagation of energy through space or material medium. However, in food preservation electromagnetic radiations (EM) are important and these are self propagating eclectic or magnetic waves or vibrations of different wavelength. Generally, radiations of shorter wavelength are more damaging to microorganisms than long wavelengths. EM radiations of importance in food preservation. The EM radiations of importance in food preservation are;

(i) Ultra violet rays
(ii) X - rays
(iii) Gamma rays
(iv) Microwaves

In food preservation ionizing radiations of wavelength of 2000 Ao or less are important. These include beta rays; Gamma rays and X - rays. Ionizing radiations ionize molecules on their path and thus destroy microorganisms without raising temperature. Killing of microorganisms in foods using electromagnetic radiations without raising temperature is termed as cold sterilization. Ionizing radiations have high energy level to cause ejection of an orbital electron from an atom or a molecule.

Radiations Types and their Characters

1. UV Rays

Ultraviolet radiations are powerful bactericidal agents. These non-ionizing radiations of < 450 nm wavelength are absorbed by proteins and nucleic acids leading to photochemical changes and subsequent cell death. The death of microorganisms results from the production of lethal mutations in nucleic acid preventing transcription and DNA replication. UV rays are bactericidal/virucidal in the wave length between 200-290 nm, but are most effective at 260 nm (2600 Ao). These have poor penetration capacity (penetrate only 0.1 mm thickness) hence their application in food industry is limited to disinfection of air, and application on food surface especially the packaging material. However, some foods are also treated with UV radiation before wrapping. UV radiation can cause rancidity in high fat products. Workers have to be protected from UV rays as it can cause burning of skin and eye disorders.

2. Beta Rays (β-particles)

These ionizing radiation are a stream of high energy electrons emitted by radioactive substances or machine generated electrons using cathode ray tubes. Beta rays have poor penetrating power but better than UV radiations. These are known to induce radioactivity in some foods under high energy sterilization conditions (at upper limit of energy level). Hence, these are of have limited application in food preservation.

3. Gamma Rays

These are uncharged electromagnetic radiations emitted from the exited nucleus of radioactive elements such as 6°Cobalt, 137Cesium *etc.* These ionizing radiations produced by the decay of radioactive isotopes are cheapest form of radiations for use in food preservation since source elements are available as byproducts of atomic waste. Gamma rays have excellent penetration power and penetrate almost anything. These can penetrate food up to a depth of 20 cm and effective as bactericidal agents.

4. X-rays

These ionizing radiations are produced by bombarding suitable metal target with high velocity electrons. These are similar to Gamma rays in their behavior.

Radiation Energy Unit

The radiation dose or amount of radiation applied for food preservation purpose is expressive by REP unit, RAD unit or GY unit.

REP unit (Roentgen – Equivalent physical)

1 REP = absorption of 83 ergs/g of matter.

Ergs are unit of energy.

RAD unit

RAD is unit of measurement of radiation.

1 RAD = absorption of 100 ergs/g of matter

1 K rad = 1000 rads

GY unit (Gray unit)

A newer and recently used unit of radiation 1 Gray = 100 rads = 11 joules/kg

1 KGy = 105 rads

Radiation Process

The radiation process given to food is of 3 types.

(a) Radappertization

(b) Radicidation

(c) Radurization

a) Radappertization

This refers to radiation process bringing about total destruction of microorganisms. It is equivalent to commercial sterility of heat processed foods, also called radiation sterilization. This process uses radiation dose of 30~40 K gray and ensures sterile product of prolonged shelflife.

b) Radicidation

This is a radiation process used for the reduction of number of viable non - spore forming pathogens which are undetectable by any standard method. This process is equivalent to pasteurization. Radiation level of 2.5~10 KGy is used in this process.

c) Radurization

It is a process of irradiation given to minimize the load of spoilage organisms thus extending the shelflife of food. This may be considered equivalent to pasteurization. This ensures enhancement of the keeping quality of a food by causing substantial reduction in the number of viable specific spoilage microorganisms. It is suitable for extending shelflife of fresh meat, sea food, fruits, vegetable *etc.* Dose used is 0.75~2.5 KGy.

Effect of Radiation on Food Quality

Irradiation helps to improve the shelflife of food, but often it can bring about undesirable changes in irradiated food caused directly by irradiation and indirectly by post irradiation reactions. The free radicals that are produced during radiation process in some foods can bring about oxidative changes which can result in product discoloration, tissue softening (in fruits) due to degradation of pectin and cellulose, and development of rancidity in high fat products due to the production of carbonyl and peroxide radicals during radiation and subsequent storage in presence of oxygen. Product discoloration and rancidity can be reduced by giving radiation process at low temperature in the absence of oxygen.

Foods Permitted for Irradiation

Irradiation is permitted in many countries for several foods. These include,

1. Inhibition of sprouting of potatoes, onion, garlic, mushrooms *etc.*
2. Decontamination of food ingredients (species), and insect disinfection in cereals and grains.
3. Destruction of parasites in meat
4. Inactivation of *Salmonella* in poultry, eggs, shrimps, frog legs.
5. Delay in fruit maturation (strawberries, mango, papaya).

6. Mould and yeast reduction in many foods
7. Radiation resistance of microorganisms

Microbicidal effect of irradiation is due to the direct interaction of radiation with key molecules within the microbial cell as well as inhibitory effect of free radicals (H+, OH-ions) produced by the radiolysis of water. Resistance to radiation varies with different microorganisms. Among microorganisms, Gram negative bacteria are more sensitive than Gram positive followed by spores. Many spoilage bacteria of seafoods are Gram negative, hence are least resistant to irradiation. *E. coli* is being highly sensitive not useful as indicator of fecal contamination in irradiated foods. Gram positive bacteria like *S. aureus, Micrococcus, Bacillus* and *Clostridium* are more resistant. Viruses are extremely resistant to irradiation.

Radiation Resistant Bacteria

Among microorganisms, *Deinococcus radiophilus*, a Gram positive, non-spore former is most resistant to radiation and can survive radiation of 15 K Gy. However, D. radiodurans is most well studied and first radio-resistant organism isolated. The actual mechanism is of resistance is not clearly known, but it is attributed to its unusual cell wall composition (absence of techoic acid), presence of outer membrane and pigmentation besides other mechanisms.

3.9 Microbiology of Value Added Products

1. Thermally Processed RTE Products

Most of the RTE aquatic products receive thermal processing at any one of the stages of processing. Retorting, cooking or pasteurization can be employed to prepare RTE products. Retorting (often referred as canning) is a heat treatment employed at 110°C or above to eliminate all the food borne pathogenic bacteria to produce a shelf stable product at ambient temperature. *Clostridium botulinum* spores is chosen as the target organism and a 12 D process is sufficient to ensure safety of canned aquatic foods (FDA, 2001). *Listeria monocytogenes* is selected as the target pathogen and a 6D process is sufficient to ensure safety of cooked products. In case the product is cooked before vacuum packaging or modified atmospheric packaging, it is necessary to eliminate the spores of *Clostridium botulinum* type E and non-proteolytic types B and F. A 6D process is designed with *Clostridium botulinum* type B as the target pathogen. FDA recommendations for cooking aquatic food products to destroy organism of public health concern in food processing operations and food services; along with cooling of cooked foods (Table 3.10).

Pasteurization is a relatively mild heat treatment carried out at 74°C to eliminate the most resistance pathogens and reduce the spoilage bacteria associated with the product. It is performed after the product is placed in the hermetically sealed container and are distributed either refrigerated or frozen. *e.g.* pasteurized crab meat, lobster meat, *etc.*

Table 3.10: FDA Recommendations for Cooking Aquatic Food Products

Task	*Recommendations*
To destroy organism of public health concern in food processing operations	A 6D reduction of Listeria monocytogenes or if necessary A 6D reduction of Clostridium botulinum type E (FDA, 2001)
To destroy organisms of public health concern in food service	Food containing raw fish shall be cooked to heat all parts at 63°C or above for 15 sec (FDA, 1999)
	Food containing comminuted fish shall be cooked to heat all parts at 68°C for 15 sec (FDA, 1999)
	Food containing stuffed fish shall be cooked to heat all parts at 74°C for 15 sec (FDA, 1999)
	Alternative cooking time/temperature combinations can be used by the food processors so long as the same lethal effect as 75°C is achieved. Scientifically accepted alternative time/temperature combinations include: 70°C for 2 min, 67°C for 5 min and 64°C for 12 min and 37 sec.
For cooling cooked aquatic food products	Cooked products should be cooled from 60°C to 21.1°C or below within 2 h and to 4.4oC or below within another 4 h (FDA, 2001)

FDA, 2001.

These products are known as "Pasteruized RTE foods". *Clostridium botulinum* type E and non-proteolytic types B and F are the target organisms for the products that are not VP or MAP but contain a barrier. A 6D process for *Clostridium botulinum* type E and non-proteolytic types B and F is sufficient to ensure safety.

2. Raw RTE Products

Consumption of raw fish or shellfish is also quite common in Japan and some South East Asian countries. Major risks associated with the consumption of raw or undercooked fish and shellfish are with *Salmonella* and *Vibro vulnificus*. *Vibro vulnificus* is a bacterium that lives in warm seawater and is not caused by pollution. Good food safety practices and proper cooking can prevent the risks. Cook the fin fish until it reaches 145°F or until the flesh is opaque and separates easily with a fork; cook shellfish until the flesh is opaque; or, cook clams, oysters, and mussels, until the shells open. Sushi is a potentially hazardous food as it contain perishable ingredients and involves manual handling. The potential risks associated with sushi are with *Listeria monocytogenes* and *Bacillus cereus*. Faecal coliforms and *E. coli* are of risk during summer. Aerobic colony counts or *E. coli* counts reflect the general hygiene status of sushi and sashmi. Microbiological guidelines are available for the RTE foods placed in the market *i.e.* supermarkets, convenience stores, food stalls, distributors, whole salers, catering establishments, manufacturers and point of import.

Table 3.11: EU Guideline Limits for Pathogens and Toxins in RTE Food Placed in the Market

Pathogen/Toxin	*cfu/g or Absence or Presence in 25 g*		
	Satisfactory	*Borderline*	*Unsatisfactory*
Bacillus cereus and other pathogenic *Bacillus* spp.	< 10^3	10^3 - <10^5	> 10^5
Camphylobacter spp.	Not detected	--	Detected
Clostridium perfringens	< 10	10 - <10^4	> 10^4
Coagulase positive staphylococci	< 20	20 - <10^4	> 10^4
Salmonella spp.	Not detected	--	Detected
Shigella spp.	Not detected	--	Detected
Staphylococcal enterotoxin	Not detected	--	Detected
Vibrio cholerae (O1 and O139)	Not detected	--	Detected
Vibrio parahaemolyticus	< 20	20 - <10^3	> 10^3

Table 3.12: FDA and EPA Safety Levels in Regulation and Guidance

Product	*Organisms*	*Level*
RTE fishery products (minimal cooking by consumer)	*Listeria monocytogenes*	Presence of organism in 25 g sample
	Vibrio cholera	Presence of toxigenic O1 or O139 or nonO1 and nonO139 in 25 g sample
	Vibrio parahaemolyticus	Levels less than 50/g (MPN)

Table 3.13: Lists Aerobic Colony Count (ACC) Limits for Various Categories of RTE Foods

Food Category	*ACC (cfu/g)*		
	Satisfactory	*Borderline*	*Unsatisfactory*
1. Ambient stable canned and pouched foods	< 10	NA	If present
2. Foods cooked immediately prior to sale or consumption	< 10^3	10^3 - <10^5	> 10^5
3. Cooked foods chilled but with minimum handling prior to sale or consumption; Canned pasteurized foods requiring refrigeration	< 10^4	10^4 - <10^5	> 10^7
4. Cooked foods chilled but with some handling prior to sale or consumption	< 10^5	10^5 - <10^7	> 10^7
5. Foods mixed with dressings, dips, pastes	< 10^6	10^6 - <10^7	> 10^7
6. Extended shelf life food products requiring refrigeration	< 10^6	10^6 - <10^8	> 10^8
7. Raw ready to eat products; cold smoked products	< 10^6	10^6 - <10^7	Yeast > 10^6 Gm (-) bacilli > 10^7 Lactics > 10^8
8. Pickled, marinated or salted foods	NA	NA	–do–
9. Dried foods	NA	NA	–do–
10. Fermented, cured and dried meats	NA	NA	–do–

Other Miscellaneous Fish Products

Prolonged shelf-life in foods has traditionally been associated with thermal processing, alone or in combination with chemical or biochemical preservation methods. However, heat processing tends to reduce product quality and freshness and also destroys the functionality and flavor of food.The ideal processing method would therefore be one that inactivates microorganisms and halts deteriorative reactions by other means than heating or freezing. Novel non-thermal processing like High Pressure Processing (HHP), Pulsed Light Processing (PLP), Irradiation, Ultrasound, Pulsed Electric Fields (PEF), Oscillating Magnetic Field (OMF), Ultraviolet Processing, Ozone Processing can render (aqua) foods free of pathogenic and spoilage organisms, retain colour and flavour, improve texture and shelf life. These processing methodssatisfy the consumers demand for minimal processed aquafoods with higher sensory quality and minimal added chemicals than the conventionally processed aquafoods.

1. High Pressure Processing (HPP)

It is also known as High Hydrostatic Pressure (HHP) or Ultra High Pressure (UHP) processing or Pascalization. HPP uses the pressure up to 900MPa to kill microorganisms found in foods, even at room temperature. When high pressures up to 1000MPa are applied to packages of food that are submerged in a liquid, the pressure is distributed instantly and uniformly throughout the food. Typically a pressure of 350MPa applied for 30min or 400MPa for 5min will cause a 10 fold reduction in vegetative cells of bacteria, yeasts or moulds. High pressure processing has no "heating or cooling" periods and there is a rapid "pressurization/ depressurization" cycle, thus reducing processing times compared to thermal processing. Normally, gram-positive vegetative bacteria are more pressure resistant (baro-tolerant) than vegetative cells of gram-negative bacteria. Two most well studied pathogenic non-spore forming gram-positive bacteria regarding the use of HPP are *Listeria monocytogenes* and *Staphyloccocus aureus*. *Staphyloccocus aureus* appears to have a high resistance to pressure. *Vibrio parahaemolyticus*is substantially more sensitive to the effects of high hydrostatic pressure than *L. monocytogenes*. HPP can inhibit microbial growth and cause cells to die. However, spores of bacteria are extremely resistant to inactivation by pressure, but can be killed at >1,200 MPa unlike vegetative types which can be killed at just at 600 MPa. Spores of *C. botulinum* are among the most pressure-resistant known. Bacterial spores are not destroyed unless high hydrostatic pressures (in excess of 800MPa) are used.Heat in conjunction with HPP is a requirement for effective elimination of bacterial endospores in low-acid foods. Yeasts are an important group of spoilage microorganisms, but none is an important food pathogen. Toxic mold growth is a safety concern in foods. Most human viruses will be eliminated in pressure treatments designed for elimination of problematic bacteria (for example, 400 MPa). Among viruses, bacteriophage are relatively easy to handle and enumerate, and would carry no risk of infection to humans.

2. Pulsed (or High Intensity) Light Technology

High intensity light is a pulsed broad spectrum white light. Pulsed light technology is a decontamination or sterilization technology that can be used for the rapid inactivation of microorganisms on food surfaces, equipment and food packaging materials. It is a non-thermal preservation intervention, with the ability to minimize the deleterious effects of thermal processing and chemical treatments on quality and sensory attributes. Pulses of light used for food processing applications typically emit one to twenty flashes per second of electromagnetic energy.

The impact of pulsed light to inactivate microorganisms isolated from fish products was studied by Hurdletech project of Seafood Plus showed that a short treatment time (325 μs) at relatively low dose induced high inactivation (> 7 Log10 CFU ml-1 or cm-2) of *L. innocua*. Pulsed light efficacy is dependent on the light dose received by microorganisms, which was modified by some process factors such as pulse energy, number of pulse. *L. innocua* and *L. monocytogenes* were the most pulsed light resistant bacteria among the different seafood spoiling and pathogenic strains studied. Therefore, *L. innocua* could be considered as a surrogate for *L. monocytogenes* and as a reference microorganism for pulsed light treatment optimization in seafood products.

The mode of action of the pulsed light process is attributed to unique effects of the high peak power and the broad-spectrum of the flash. A primary cellular target is nucleic acids. Inactivation occurs by several mechanisms, including chemical modifications and cleavage of the DNA. The impact of pulsed light on proteins, membranes, and other cellular material probably occurs concurrently with the nucleic acid destruction. For example, the motility of *E. coli* ceases immediately after exposure to pulsed light. Doses for sterilization by pulsed white light are an order of magnitude lower than that for UV exposure. *Bacillus subtilis*, for example is sterilized (99.99 per cent disinfection) by about 42,600 microW-s/cm2 of UV while requiring a dose of only 4500 microW-s/cm2 under pulsed light.

Role of Antimicrobials

Fish preservation using antimicrobials is one of the common method of preservation. Antimicrobials are the compounds which inhibit the growth or kill the microorganisms like bacteria, fungi, molds. Antimicrobials are often used to extend the shelf life of fish by reducing microbial proliferation during transportation, processing and storage. Growth of bacteria and spoilage of fish are dependent on the species of bacteria, nutrients availability, pH, temperature, moisture and gaseous atmosphere. Antimicrobial compounds added during processing should not be used as a substitute for poor processing conditions or to cover up an already spoiled product. Antimicrobials offer a good protection for fish in combination with refrigeration. Common antimicrobials in fish preservation include: chlorides, nitrites, sulfides and organic acids.

Classification

Antimicrobials used to eliminate microbial pathogens are either of biological or chemical origin. Biological antimicrobials refer to those that employ living organisms or agents derived from them. Chemical agents are chemicals employed to eliminate pathogens.

Chemical Preservatives

Chemical preservatives includes chlorine related compounds, acids, nitrites and sulphites. Chlorine and its related compounds have strong antimicrobial properties. Choloro-acids are widely used to clean food contact surfaces and raw material in the food industry. Chloro-acids such as sodium hypochlorite can be easily generated by electrolyzing dilute salt solutions (0-1 per cent NaCl) using an anode and a cathode. Sodium hypochlorite solution generated from this process are referred to as "electrolyzed water". Electrolyzed water exhibits strong antimicrobial action against *V. parahemolyticus* and *V. vulnificus* in oysters treated with 30 ppm. Chlorine affects the cell membrane and cellular DNA resulting in cell death.

Sodium Chloride

Sodium chloride (NaCl) has a long history of use in food preservation as an additive as well as preservative in sufficiently high concentrations. It inhibits microbial growth by increasing osmotic pressure as well as decreasing the water activity in the micro-environment. A concentration of 20 per cent of NaCl is enough to inhibit many food spoilage yeasts and some bacteria can be inhibited by concentrations as low as 2 per cent. However, some microorganisms have shown ability to tolerate high concentrations of NaCl such as Micrococci and Bacillus.

Nitrites

The nitrites used in fish preservation industry are always in the form of salts such as sodium nitrite or potassium nitrite. They are long known as antimicrobials preventing the growth of the toxin producing *Clostridium botulinum, Staphylococcus aureus* and *Yersinia enterocolitica*, which would grow under anaerobic environment in vacuum packages. Nitrite salts are effective in stabilizing the redmeat color, retaining the cured meat flavor and controlling the lipid oxidation in addition to controlling the anaerobic bacteria.

Nitrites affect the growth of microorganisms in food through several reactions including:

- ✰ Reaction with alpha-amino groups of the amino acids at low pH levels,
- ✰ Blockage of sulfhydryl groups which interferes with sulfur nutrition of the organism,
- ✰ Reaction with iron-containing compounds which restricts the use of iron by bacteria,
- ✰ Interference with membrane permeability which limits the transport across cells.

Sodium nitrite at 200 mg/kg and pH of 6.0 retarded the growth of *Achromobacter, Aerobacter, Escherichia, Flavobacterium, Micrococcus* and *Pseudomonas* species in meat. The germination of *Clostridium sporogenes* spores in meat was impended with sodium nitrite. The toxin production by *Clostridium botulinum* was prevented at 200 µg nitrite/kg of meat. The current limit for nitrite in food is 156 ppm in US, and 200 ppm in Canada for meat products. On the other hand, the use of nitrite as food additive may form carcinogenic nitrosamines under prolonged exposure. However there is no epidemiological evidence to support the relationship between nitrite consumption and a specific cancer or cancer risk.

Sulphites

As antimicrobial agent, sodium sulfite is efficient against aerobic gram-negative bacilli, molds and yeasts in fish and fishery products. Sulphites were used as antimicrobial agents in specified comminuted products such as fresh sausage because of their efficacy in controlling Enterobacteriaceae including pathogenic *Salmonella*. The antimicrobial activity is the result of the undissociated sulfurous acid, which enters the cell and reacts with thiol groups of proteins, enzymes and cofactors. Yeast cells are attacked by sulfite because sulfite reacts with cellular adenosine tri phosphate (ATP) and blocks the cystine disulfide linkages. Sulphur dioxide and their salts potassium bisulphite, potassium metabisulphite, sodium bisulphite, sodium metabisulphite and sodium sulphite collectively known as sulfites are removed from GRAS (Genarally Regarded As Safe) list and they are not allowed for use as preservative in meat in the U.S. and Canada because of the degradation of vitamin, thiamine by sulphites.

Acids

Acid antimicrobials are useful agents for controlling pathogenic bacteria in seafood and include lactate, acetate and citrate. Lactic acid inhibits *V. vulnificus* associated with raw oyster at 300 ppm. Acetate and its salt prevent the outgrowth of *C. botulinum* in fermented sushi. Acetate is cheap, widely available and well accepted by consumers. Citrate is weaker than acetate but has pleasant sour taste and control *C. botulinum* growth and toxin production in shrimp puree.

Ascorbic Acid

Ascorbic acid (vitamin C), sodium ascorbate and D-isoascorbate (erythorbate) are used as fish preservatives. Their antioxidant properties can oxidize reactive oxygen species producing water. Ascorbic acid has been shown to enhance antimicrobial activity of sulfites and nitrites The enhanced activities include both the antioxidant properties and the sequestering of iron via reduced nitrosamine formation at a level of 550 ppm, when they were used in combination with nitrite. The addition of 500 mg sodium erythorbate/kg of meat and 200 mg sodium nitrite/kg of meat had prevented the growth of *Bacillus cereus* spores in sausages kept at 20°C for 48 h.

Benzoates

Benzoic acid and sodium benzoate are used as preservatives in the fish industry. The undissociated molecule of benzoic acid is responsible for its antibacterial activity. The benzoic acid is generally used to inhibit yeasts and fungi rather than bacteria. The benzoic acid, in their unassociated state under low pH (2.5-5.0) readily crosses the cell membrane. Once entered the cytosol, the acid dissociates because of neutral pH environment. The dissociated molecules (anion and cations) cannot diffuse back across the membrane and hence accumulate in the cytosol. The acidification of the cytosol and the depletion of ATP will cause physiological malfunction and finally inhibition of microbial growth.

The effect of sodium benzoate (0.08-0.25 per cent) on the growth of *Listeria monocytogenes* in ready-to-eat meat products over 18 weeks storage at 4°C, examined showed that under low moisture content concentration of 0.1 per cent sodium benzoate was effective in inhibiting *Listeria monocytogenes.* Some of the food spoilage yeasts such as Saccharomyces and Zygosaccharomyces have intrinsic ability to resist benzoic acid and its salts. Zygosaccharomyces bailii and Yarrowia lipolytica are the most resistant yeast to benzoic acid at pH 5.0. The combination of benzoic acid treatment and nitrogen starvation conditions was suggested to enforce effective food preservation from yeast spoilage. In Canada, benzoic acid is not permitted for unstandarized meat and the limit for benzoic acid is 1000 ppm for marinated or cold-processed packaged meat. In the United States, the maximum permitted level for benzoic acid and sodium benzoate is 0.1 per cent as GRAS preservatives.

Sorbates

Sorbic acid (2, 4-hexadienoic) and its salts are widely used as fish preservatives for inhibiting bacteria and fungi. A concentration of 0.3 per cent sorbates in food is high enough to inhibit the microorganisms. The sorbic acid has an inhibitory mechanism via depression of internal pH and it may interfere with the bacterial spore germination followed by inhibiting the activity of several enzyme systems and interfering with substrate and electron transport mechanisms. Sorbic acid may act more like membrane-active substance rather than as a weak-acid preservative. Sorbate (0.1 per cent w/w) has inhibitory effect on the growth of *Clostridium botulinum* on cooked uncured sausage at 27°C. Spray of potassium sorbate (2.5 per cent), sodium acetate (2.5 per cent), sodium citrate (2.5 per cent) inhibited *Bacillus* spp. to a minimum and extended the lag phase of all organisms including psychrotrophs (*Pseudomonas*) throughout the refrigerated storage of fish at 5-7°C. A solution of 10 per cent potassium sorbate can be used in uncooked and dried sausages, because they are effective inhibitor of mold, yeast and some highly aerobic bacteria in meat. According to the Canadian Food and Drug Act, the allowable limit of potassium sorbate is 1000 ppm.

Table 3.14: List of Commercially Important Antimicrobials as Fish Preservatives

Antimicrobials	Products	Maximum Permitted level (MPL)
Acetic Acid	Preserved fish	2500 ppm
Ascorbic Acid	Canned tuna	150 ppm
	Frozen minced fish	
Calcium Ascorbate	Canned tuna	150 ppm
	Frozen minced fish	
Sodium Erythorbate	Canned clams	350 ppm
Benzoic Acid	Fish	1000 ppm
Propyl-ρ-hydroxy Benzoate	Fish	1000 ppm
Propyl Paraben	Fish	1000 ppm
Sodium Diacetate	Preserved fish	0.25 per cent of final product weight
Sodium Metabisulphite	Canned sea snails	Not exceed 100 ppm
Sulphurous Acid	Fish	500 ppm
Sulphurous Acid	Crustaceans	Not exceed 100 ppm

Biological Preservatives

Antimicrobials can be classified as biological, when it is obtained from raw materials of vegetable, fruit, herbs/spices or microbiological origin. Living organisms possess defence antimicrobial mechanism, which is exploited to control pathogens in aquatic foods.

Plant based Antimicrobials

Antimicrobials of plant origin includes alkaloids, flavanoids, isoflavonoids, tannins coumarins, glycosides, terpenes and phenolic compounds. Phytochemicals in plants are broadly grouped into phenolic compounds, terpenoids and essential oils, alkaloids, lectins and polypeptides. The phenolic compounds include simple phenols and phenolic acids, quinones, flavonoids and tannins. Phenolic compounds are secondary metabolites and one of the most widely occurring phytochemicals in plants. They contribute to the sensory properties when added to food and have antioxidant and antimicrobial properties, characteristics in extending the shelf-life of food. Other classes of compounds such as polyamines, glucosinolates and glucosides have potential as natural antimicrobials and glucosinolates, in particular are identified for antifungal, antibacterial and antioxidant properties, in addition to other functional properties. Allyl isothiocyanate (AITC), a hydrolysis product of glucosinolates of horseradish is a potent antibacterial compound and is used as a preservative in the food industry.

Plant metabolites causes their antimicrobial effect by cell membrane perturbation leading to cell dysfunction. Some metabolites such as phenols may interfere with germination enzymes and L- alanine utilization that affects germination of spores of pathogenic bacteria. AITC impregnated labels are currently used to control the growth of *Vibrio* spp. and other microorganism in raw seafood lunch boxes in Japan. Phytochemicals, thymol and gallic acid prevent the growth of *C. botulinum*, mint, thymol and garlic inhibit the growth of *L. monocytogenes*; and thyme, bay leaf and mint, inhibit the growth of *V. parahemolyticus*.

Wood Smoke

Smoking is one of the oldest method used for flavor enhancement as well as for preservation. Wood smoke contains a large number of volatile compounds that may have bacteriostatic and bactericidal effect. Formaldehyde is considered most effective among these compounds, followed by phenols and cresols. Other compounds in the smoke are aliphatic acids, primary and secondary alcohols, ketones, acetaldehyde and other aldehydes, waxes, resins, guaiacol and its methyl and propyl isomers, catechol, methyl catechol, pyrogallol and methyl ester. The residual effect of smoke in the food has been reported to be greater against bacteria than against molds.

Biopreservation

Bacteria have been used to control pathogens in food. Whole cultures as well as peptides and/or metabolites are useful in the inhibition of pathogens. Lactic acid bacteria (LAB) and/or their metabolites are used in bio preservation for their antimicrobial properties. Important agents are probiotics, nisin, pediocin, reduterin and sakacin. *Lactobacillus plantarum* and other lactic acid bacteria play an important role in the control of pathogens during seafood fermentation through acidification and production of alcohols. Nisin, produced by *Lactococcus lactis* is effective against *L. monocytogenes* and *C. botulinum*. Nisin was the first bacteriocin used on commercial scale in food. It eliminates the psychrotrophic seafood pathogen *L. monocytogenes* in fresh and lightly preserved seafood. Toxin production by *C. botulinum* type E spores in smoked mackerel can be delayed by spraying with nisin before packing in CO_2 atmospheres. Sakacin P and/or L. sake cultures (sakacin P producer) showed inhibitory effect against *L. monocytogenes* in cold-smoked salmon.

Bacillus spp. have an antimicrobial action against gram positive and gram negative bacteria, as well as fungi, and can be used as a potential biopreservative in food processing due to its wide antimicrobial spectra. Bacteriocins are peptides or complex proteins produced by lactic acid bacteria (LAB) that are active against other bacteria, principally closely related species. Bacteriocin production in *Bacillus* spp. include Subtilin from *B. subtilis*, Megacin from *B. megaterium* and Thermacin from *B. stearothermophilus*. Bacteriocins and protective cultures are effective in controlling the growth of *L. monocytogenes* in vacuum-packed cold smoked salmon.

Biocontrol

A nonbacteriocin-producing strain of *C. piscicola* was as effective in the inhibition of *L. monocytogenes* in vacuum-packed cold-smoked salmon. The growth inhibition of *L. monocytogenes* was probably due to the competitive growth of *C. piscicola* that resulted in depletion of essential nutrients. Nisin in combination with carbon dioxide and low temperature reduced the survival of *L. monocytogenes* in cold-smoked salmon. Nisin Z, carnocin UI49, and a preparation of crude, bavaricin A caused shelflife extension of brined shrimps. In vacuum-packed cold-smoked rainbow trout, nisin, sodium lactate, or their combination have inhibited *L. monocytogenes* and mesophilic aerobic bacteria. Biopreservation may be effectively used in combination with other preservative factors (called hurdles) to inhibit microbial growth and achieve food safety. The use of an adequate mix of hurdles is not only economically attractive; it also serves to improve microbial stability and safety, as well as the sensory and nutritional qualities of a food.

Chapter 4

Food-Borne Pathogens in Fish and Fishery Products

Study of food-borne pathogens involved in infective and intoxication type of food poisoning

Foodborne Pathogens

Several human pathogenic microorganisms are associated with food and water in their natural environment, and foods also get contaminated with pathogens during handling and processing. Thus, ingestion of food or water contaminated with pathogens leads to food poisoning affecting the health of consumers.

Food Poisoning

Food poisoning is an illness caused either by the ingestion of toxin elaborated by the pathogens or from the infection of the host through the intestinal tract due to ingestion of live organisms.Food poisoning by bacteria can be of two types;

- ☆ Bacterial food intoxication
- ☆ Bacterial food infections

Bacterial Food Intoxication

It is a food borne illness caused by the presence of a bacterial toxin formed in the food.

It is caused due to consumption of food containing pre- formed toxin but not live bacteria.

Two types of food intoxications are caused by bacteria.

1. **Botulism:** Food poisoning caused by the ingestion of food containing toxin produced by *Clostridium botulinum*.
2. **Sataphlococcal intoxication:** Food poisoning caused by the ingestion of food containing pre-formed toxins produced by Staphyllococcus aureus

Bacterial Food Infections

This is a food borne illness caused by the entrance of live bacteria into the body through ingestion of contaminated food. The illness mainly results from the reaction of the body to pathogen's presence or their metabolites.

Ex: *Salmonellosis*, *C. perfringes* illness, Cholera, *V. parahaemolyticus* infection *etc.*

Sources of Pathogens

Seafood/foods of aquatic origin have potential to cause human illness. The entry of bacterial pathogens to aquatic environment/consumers results from the following:

1. Pollution of aquatic environment from sewage or animal wastes.

- ☆ Consumption of raw or semi processed contaminated food/water.
- ☆ Infected domestic animals used for food. (Ex: *Salmonella* in poultry)
- ☆ Unsanitary practices of product handling through food handlers.
- ☆ Presence of naturally associated human pathogens in aquatic foods/ environment

Natural Habitat of Pathogen	*Infection*	*Intoxication*
Aquatic environment	*Vibrio* spp. *Aeromonas* *Plesiomonas*	*C. botulinum* Type E (non-proteolytic)
General environment	*Listeria monocytogenes*	*C. botulinum* Type A,B (proteolytic) *C. perfringens* *Bacillus cereus*
Animal-human reservoir	*Salmonella* *E. coli* (EPEC, ETEC) (EHEC) *Shigella* *Campylobacter*	*Staphylococcus aureus*

4.1. Indigenous Pathogens

Human Pathogenic Vibrios

Members of the Vibrionaceae family occurring in marine environment are responsible for many of the reported cases of infection worldwide. Since majority of the species are inhabitant of the marine aquatic environment, the ecological niche of the region is liable to alter with the bacterial abundance. The occurrence of these organisms eventually cause contamination of seafood, which harbor pathogens and are transmitted through foods and through cross contamination during handling and processing of foods. Consequently, it is expected that most of the consumers are constantly exposed to the risk of contracting disease from these pathogens. The genus Vibrio includes more than 90 species, of which, 12 species have been identified to as human pathogens. The most important human pathogenic *Vibrio* includes *Vibrio cholerae, V. parahaemolyticus* and *V. vulnificus* and their infection follows either direct contact with aquatic environment or indirectly via contaminated food and water.

Vibrios are more common in warm coastal waters. There are some vibrios that require only small amounts of Na+ for growth such as *Vibrio cholerae* and *V. mimicus* and they are present in freshwater rivers and lakes. In marine environments, vibrios are isolated from sediment, water and aquatic plants and animals, more commonly from filter feeding shellfish, such as oysters. Plankton represents an organically rich microenviroment for vibrios.Vibrios are commonly reported to cause about 28 per cent of the seafood associated outbreaks. *V. parahaemolyticus* cause more outbreaks and illnesses (24 per cent) than other pathogens like *V. cholerae* non-01, *V. cholerae* non-0139 and *V. vulnificus.*

Incidence of Vibrios in Aquatic Foods

Vibrios are associated with a great variety of aquatic foods. About 36-60 per cent of fishfish and shell fish are contaminated with *Vibrio* spp. especially with *V. parahemolyticus* and *V. alginolyticus.* particularly greater during the summer months. The order of prevalence of vibrios in seafood: *V. alginolyticus* (81 per cent), *V. parahaemolyticus* (77 per cent), *V. cholerae* non-01 (31 per cent), *V. fluvialis* (27 per cent), *V. furnishsii* (19 per cent), *V. mimicus* (12 per cent) and *V. vulnificus* (12 per cent).

1. *Vibrio cholerae*

Vibrio cholera is a causative organism of human cholera due to the consumption of contaminated food/water containing viable cells. *V. cholera* is a Gram negative, non spore forming, facultatively anaerobic, curved rods belonging to the family enterobacteriacea. The epidemic/pandemic cholera is caused by *V. cholera* serovar “O” Group 1 (01) and consists of 2 biotypes *viz.* classical and El-Tor based on biochemical tests. These have ability to agglutinate 01 antiserum. Strains that do not aglutinate in 01 antiserum are referred to as non 01 or non-agglutinating vibrios (NAGs). The non 01 strains are autochthonous to estuarine environment and are widely distributed. These are generally non-pathogenic but known to cause gastroenteritis, soft tissue infection and septicemia in humans. *V. cholerae* grow well at 37°C, tolerant to alkali (pH up to 9.2), survive for long period in fresh and marine environment and enter in to VBNC phase during adverse conditions. Infections occur through oral route by contaminated food and water. Attach to intestinal mucosa and produce enterotoxin and cause loss of water and electrolytes due to the disruption in the cyclic AMP system.

Symptoms of Illness

Symptoms vary from mild to severe diarrhea (rice water stool), abdominal cramps, nausea, vomiting, dehydration, shock and death in severe cases. Infection of non 01 serotype is associated with exposure to natural aquatic environments, consumption of seafood and exposure to polluted water. Non 01 *V. cholerae* show better survival and multiplication in a wide range of foods than 01 strains. *V. cholerae* 01 is the causative agent of cholera but implicated in very few food-borne diseases. There are two serogroups, 01 and 0139 that produce cholerae enterotoxin (cholera toxin or Ctx) associated with epidemic cholera. There are some strains of serogroups, other than 01 and 0139, that are pathogenic, either by the production of Ctx or other virulence factors; but do not cause epidemics or pandemics.

Classification of *V. cholerae*

V. cholerae 01 serogroup can be sub-divided into serotypes called Ogawa, Inaba and Higojima. Based on lipopolysaccharide (LPS) somatic antigen, *V. cholerae* 01 can also be divided into two biotypes: Classical and El Tor. El Tor strains are associated more with asymptomatic infections and few fatalities, whereas Classical strains cause severe clinical manifestations. Cholera toxin is conserved among these strains but sequence differences exist in CtxB epityping and ctxB genotyping. A new epidemic emerged from Eastern Indian and Bangladesh in 1993, revealed the bacterium belonging to a new serogroup 0139 (Bengal) strain. This strain is indistinguishable from typical EL Tor *V. cholerae* 01 strains, but do not produce 01 LPS.

All strains negative for 01 serogroup are referred to as "non 01 *V. cholerae*" or "non cholerae vibrios (NCV)" or "non agglutinable (NAG)" vibrios. These strains are isolated from cases of diarrhea and extra intestinal infections. Most of these strains do not produce Ctx, but produces polysaccharide capsule. However, strains of 0141 serogroup produce Ctx and toxin coregulated pilus (TCP) colonization factor, typical of 01 and 0139 strains and also produce type III secretion system (T3SS). The *V. cholera* 0139 also produces a polysaccharide capsule like non 01 *V. cholerae*.

Natural Habitats

V. cholerae is part of the normal free living bacterial flora in estuarine waters. Non-01/0139 strains are commonly isolated from the environment, which are almost Ctx negative.

V. cholerae are capable of colonizing the surface of zooplanktons such as copepods (100 cells in a single copepod). *V. cholerae* produces a chitinase enzyme, that binds to chitin, the principal component of crustacean shells. Chitin provides food, adaptation to environmental nutrient gradients, tolerance to stress, and protection from predators.

Peristence of *V. cholerae* within the environment facilitate them to assume survival forms, such as "viable but non culturable state", biofilms, and a rugose survival form. The VBNC state is a dormant state with reduced cell size (ovoid) and can be assessed by direct viable count procedures. For the formation of biofilms, factors including flagella, type IV pilus mannose sensitive hemagglutinins (MSHA), and quorum sensing (QS) are involved.

Biofilms allow the bacterium to persist in association with biotic and abiotic surfaces and help to prevent predation by grazing protozoa. *V. cholerae* cells present in biofilms are much more resistant to killing by acid shock than are planktonic cells, which help them to survive stomach acidity after ingestion by human. The formation of "rugose" or "wrinkled" colonial morphology is due to the production of exopolysaccharide to protect against adverse environmental conditions. Rugose variants survive in the presence of chlorine and other disinfectants. The rugose polysaccharide produced by O1 El Tor strain is called "VPS" and genes required for the synthesis are clustered in vps locus.

Disease Characteristics

Potentially fatal dehydrating diarrhea is characteristic of cholera in persons infected with *V. cholerae* O1/O139. About 11 per cent of patients are with classical infections and 2 per cent with El Tor infections. The incubation period range from several hours to 5 days. The onset of illness is sudden with profuse watery diarrhea or there may be anorexia, abdominal discomfort, and simple diarrhea. Gastroenteritis associated with *V. cholerae* non-O1/O139 is generally of mild to moderate severity. Non bloody and occasionally bloody diarrhea occurs, besides abdominal cramps and fever with nausea and vomiting. Non-O1/O139 is frequently isolated from extra intestinal infections such as septicemia, wound infections and ear infections due to exposure of fresh or brackish water. The case fatality rate can exceed 50 per cent and pose more risk to individuals with liver disease. A dose of 1011 cells of *V. cholerae* is required to cause diarrhea.

Virulence Mechanism

Infection due to *V. cholerae* O1/O139 begins with the ingestion of food or water contaminated with the bacterium. After passage through the acid barrier of the stomach, vibrios colonize the epithelium of the small intestine by means of one or more adherence factors. Invasion into epithelial cells (lamina propria) does not occur. Production of cholera enterotoxin disrupts ion transport by intestinal epithelial cells leading to subsequent loss of water and electrolytes, causing severe diarrhea.

Cholera is induced by cholera enterotoxin also referred to as cholera toxin (Ctx).

Zonula occludens toxin (Zot) that increases the permeability of the small intestinal mucosa by affecting the structure of the intercellular tight junctions (zonula occludens).

Accessory cholerae enterotoxin (Ace) that causes fluid accumulation in intestinal loops and increases potential difference in intestinal tissue.

Rtx toxin is produced by *V. cholerae* 01 El Tor and 0138 strains and is related to the members of the RTX (repeat in toxin) toxin family, a group that includes the hemolysin of uropathogenic.

Colonization Factors

Toxic coagulated pilus (TCP) is the best characterised intestinal colonization factor of *V. cholerae*. The TCP pili belonging to the type IV family consists of long filments of 7 nm in diameter that are laterally associated in bundles. The role of TCP is to mediate interbacterial aggregation and faciliatate intestinal colonization.

V. cholerae enter the mucus gel overlaying the intestinal epithelium. Down regulation of chemotaxis enhances initial colonization specially in the upper small intestine, after that, chemotaxis is upregulated to allow penetration through the mucus. *V. cholerae* flagellins can induce interkeukin 8 production in intestinal epithelial cells.

Conditions for Outbreaks

- ☆ Consumption of live cells orally through food/water.
- ☆ Consumption of post process contaminated food.
- ☆ Consumption of food without heating before serving.
- ☆ Consumption of contaminated food/water.

Prevention of Outbreaks

- ☆ Avoiding consumption of contaminated food and water.
- ☆ Preventing contamination by following sanitary measures.
- ☆ Keeping the surroundings of processing plant clean and disinfected.
- ☆ Adherence to good personal hygiene by food handlers.
- ☆ Use of potable water after chlorination.

2. *V. vulnificus*

Among the pathogenic strains, *Vibro vulnificus* is one of the serious disease causing agents responsible for 95 per cent of all deaths caused by seafood borne pathogens. Primary septicemia results due to consumption of raw oyster resulting in 50-60 per cent fatality rates. This is the highest death rate of any food borne disease agent. *V. vulnificus* infection results from consumption of raw oyster during warm months and result in primary septicemias with 96 per cent incidence. This bacterium is unusual as it produces wound as well as food infection.

Classification

Originally known a "lactose positive vibrios" and renamed as *V. vulnificus* after several phenotypic and genotypic studies. There are three biotypes. Biotype 2 is the major source of fatalities. *V. vulnificus* is susceptible to freezing, low temperatures, pasteurization, high hydrostatic pressure and ionizing radiations. There are two distinct genotypes of *V. vulnificus* based on 16S rRNA analysis. "C" genotype strains (90 per cent) are from clinical samples and "E" genotypes strains (93 per cent) are from environmental sources. They occur in equal numbers in estuarine waters. Oysters contain 84 per cent of "E" genotypes.

Natural Habitats

V. vulnificus is widespread in estuarine environments around the world. They are isolated from oysters, clams, crabs, ark shells, planktons and seawater samples. *V. vulnificus* from the intestinal tract of various bottom feeding coastal fish suggested that these fishes are the major reservoirs. A strong correlation is observed between temperature and *V. vulnificus*, as they die-off during cold weather months.

Disease Characteristics

The incubation period for primary septicemica ranges from 7 h to 10 days. The significant symptoms include fever, chills, nausea, hypotension and tolic pressure. The gastroentritical symptoms are not common. The unusual symptom that occurs is secondary lesions on the legs that frequently develop into necrotizing fasciitis or vasculitis and oftern necessitate surgical debridement or limb amputation of the affected tissues. The time of death in the fatal cases varies from 2 h to 6 weeks. Symptoms during septicemia include fever, tissue edema, hemorrhage and hypotension all associated with entotoxic shock.

Virulence Mechanism

All strains of *V. vulnificus* produce a polysaccharide capsule coat that initiates infection. Virulence strains have "opaque" (encapsulated) colony type, whereas isogenic cells have "translucent" (acapsular) colony type. Encapsulated cells are able to utilize transferring bound iron. There are 10 distinct serotypes of *V. vulnificus* associated with human infection. Elevated level of serum iron is essential for *V. vulnificus* to multiply in the human host and cause septicemia. *V. vulnificus* although produces both hydroxymate and phenolate siderophores, it is unable to compete with serum transferring for iron. Another product of *V. vulnificus* is the endotoxic LPS present in the cells, which induce nitric oxide synthase to release nitric oxide, which subsequently cause host tissue damage.

This bacterium also produces a large number of extracellular compounds including hemolysin, protease, elastase, collagenase, DNase, lipase, phospholipase, mucinase, chondroitn sulfatase, hyaluronidase and fibrinolysin. A powerful heat stable hemolysin/cytotoxic encoded by vvhA gene, possesses cytolytic activity against CHO cells and vascular permeability activity. The production of this hemolysin, a metalloprotease is regulated by a transmembrane virulence regulator homologous to ToxRS of *V. cholerae*. It enhances vascular permeability through the release of bradykinin. They can degrade type IV collagen and destroy the basal membrane layer of capillary vessels. LDL inactivates this cytolysin.

A elastolytic preotease of *V. vulnificus* can degrade albumin, immunoglobin G, elastin and complement factors. This causes extensive hemorrhagic necrosis, edema and muscle tissue destruction. A broad specificity metalloprotease, a product of the vvpE gene cleaves several plasma proteins and interferes with blood clotting functions. An RTX toxin causes rearrangement of the cytoskeletal structure and cause cellular necrosis. Proteins involved in chemotaxis of *V. vulnificus* are also important virulence gene products.

Control Measures

Control strategies include consumer education, harvest area monitoring and post harvest reduction treatments. Consumers should avoid eating raw or under cooked shellfish. Microbiological sampling of harvest area waters for Vibios is essential. The most common processing practice recommended is depuration, which relies on the mollusks inherent ability to clean itself of contamination in pathogen free water. High pressure treatment, irradiation, quick freezing and pasteurization also make oysters safer.

Vibrios are sensitive to cold temperatures, but some vibrios particularly V.parahemolyticus survive in frozen seafoods held at -30°C. *V. parahaemolyticus* cells enter into the "viable but nonculturable state (VBNC)" occurs when exposed to temperatures below 10°C. *V. vulnificus* failed to grow in oysters held at 13°C and IQF technology reduces *V. vulnificus* to non detectable levels.

All the vibrios are sensitive to heat. Heating of shellfish to an internal temperature of 60°C for several minutes is sufficient to kill them. The FDA recommends steaming of shellstock oysters, clams and mussels for 4-9 min, frying shucked oysters for 10 min at 375°C or baking oysters for 10 min at 450°C to control vibrios

Doses of 3kGy of gamma radiation kill vibrios in frozen shrimps. Hydrostatic pressure treatment (30,000 to 50,000 psi) caused 6 log reduction of *V. vulnificus* within 10 min and 9 log reduction of *V. parahaemolyticus* within 30 sec. *V. vulnificus* is inactivated by fruit or vegetable juices and spice extracts. *V. parahaemolyticus* is sensitive to 50 ppm BHA and inhibited by 0.1 per cent sorbate. Vibrios are highly sensitive to acids, at pH4.8.

3. *Vibrio parahaemolyticus*

Characters

Food poisoning by *V. parahaemolyticus* is caused mainly by the consumption of seafoods containing live bacteria. It is an autochthonous marine bacterium associated naturally with marine organisms.V. parahaemolytixcus is G-ve, straight/ curved rod, halophilic requiring 1~3 per cent salt for growth, can grow at 7 per cent NaCl, optimum temperature for growth is 35-37°C but can grow over a range of temperature (10-44°C), minimum temperature for growth is 7°C, grow in a pH range of 5~11, sensitive to freezing and chilling temperature and killed by heating at 100°C. Isolates from cases of food poisoning (about 97 per cent) are haemolytic (Kanagawa positive), and majority of environmental isolates are non-haemolytic (Kanagawa negative). Food poisoning by this bacterium causes gastroenteritis. These are killed by heating. Following good sanitation and GMP would prevent its build up in foods. This organism is the major cause of gastroenteritis in Japan.

Symptoms of Illness

Raw food of marine origin, marine fish, shellfish, crustaceans and fish products are involved in cases of food poisoning. Permissible limit in foods is 104/g. Symptoms of illness include abdominal cramps, diarrhea, nausea, vomiting, bloody stools, mild fever, chills and headache. Incubation period is 4-48 hrs and infected individuals recover with in 2-5 days.

These findings showed that virulence determinants vary from strain to strain and their complex genome flexibility may affect their fitness to the host and their infection capabilities and are recognized as a potential threat to biosecurity, public health and food security even though they are not carrying any virulence attributes which is found in human pathogenic strains. Most of the vibrios exhibit considerable strain to strain variation in terms of its virulence. It is hypothesised that several environmental and/or host-derived factors that affect transcription and expression of virulence genes. So far, no single research work has been carried out on the ecology and epidemiological surveillance of *V. parahaemolyticus*, molecular mechanism of pathogenesis of this pathogen among cultured aquatic animals. However, as long as it remains unclear, development of intervention strategies to control the incidence or risk due to the activity of such strains of *V. parahaemolyticus* will not be possible.

Prevention of Outbreak

- ☆ Cooking food thoroughly
- ☆ Chilling food rapidly
- ☆ Preventing cross contamination from marine fish
- ☆ Following sanitary measures

4. *Listeria monocytogenes*

Among Listeria species, *Listeria monocytogenes* is most important human pathogen causing the disease listeriosis and is of great concern to special risk groups. The susceptible groups include pregnant women and their fetuses, cancer patients and others undergoing immuno-suppressive therapy, as well as diabetics and cirrhotics and the elderly. Although the risk of contracting listeriosis is less for normal, healthy individuals, they may also contract the disease.

Characters

Listeria monocytogenes is a facultative intracellular pathogen. The organism enters the body through the intestine and has a variable incubation period from 1 day to a month or longer. The ingested cells enter the body through ileal villi cells, subsequently taken up by macrophage cells in the bloodstream, multiply inside the host cell and released after bursting of macrophages and liberated cells infect other cells. Listeria is widely distributed in the environment, human beings and a variety of animals including seagulls. Most of these environmental strains are non-pathogenic. *Listeria* sp. other than *L. monocytogenes* appears to be more common in tropical areas. Though it has been isolated from a variety of seafoods including refrigerated and frozen crabmeat, reports of listeriosis involving seafood is rare. Listeria are relatively heat resistant but unable to surive in foods receiving adequate heat treatment. The presence of *L. monocytogenes* in cooked seafood is mainly due to cross-contamination of the product or under processing.

Listeria are aerobic under most circumstances but can be facultatively anaerobic and thus grow well under reduced levels of oxygen in packaged products. Listeria can also survive and grow under refrigeration temperature and also survive freezing conditions.

Contaminated food is increasingly recognized as an important vehicle of L.monocytogenes. Dairy products, salads and vegetables have been implicated in outbreaks of listeriosis. Frequent isolations from seafood as well as its ability to grow in in chilled smoked salmon at +4°C is an indication of possible involvement of seafoodsin the transmission of *L. monocytogenes.*

Listeria Species Involved in Food Poisoning

Six species of Listeria are currently recognized, but only three species, *L. monocytogenes, L. ivanovii* and L seeligeri are associated with disease in humans and/or animals. However, human cases involving *L. ivanovii* and *L. seeligeri* are extremely rare.

Listeria are commonly identified by serotyping. Types 1-7 are known, with Types 1/a, 1/b, and 4b predominating as both environmental and clinical isolates. The serotyping scheme is based on both somatic and flagellar antigens. Phage typing has also been employed as a method of further identifying isolated strains. *L. monocytogenes* is subdivided into 13 serovars on the basis of somatic (O) and flagellar (H) antigens.

Symptoms

Typical Listeria infections result in septicemia, meningitis and encephalitis, and enteritis. High mortality is reported among infected individuals. Mortality rate of 29 per cent is noticed among patients in a New England outbreak involving fluid milk. This causes the transitory flu-like symptoms at the initial stage of infection with or without symptoms of stomach disturbances and diarrhea. The actual disease known as listeriosis occurs only after the attainment of severe form of septicemia, encephalitis, lesions, or meningitis. All of these forms of listeriosis is noticed among individuals who are not immuno competent.

Prevention of outbreak

To prevent risk due to *L. monocytogenes*, FDA in the US has recommended complete absence in ready-to-eat seafood products such as crabmeat or smoked fish. This zero-tolerance is also required for products which receive a listericidal treatment, and for products which have been directly implicated in a food borne outbreak. However this restriction does not apply to raw product that will be cooked before eating. Low number of *L. monocytogenes* is allowed in other types of products in which the organism is shown to die-off. But, numbers greater than 10/g are likely to constitute a risk to human beings particularly predisposed persons (very old, very young or immuno-suppressed). Following proper GMP and factory hygiene is expected to maintain the level of *L. monocytogenes* contamination on fish products at very low level of less than 1-10/g.

5. *Clostridium botulinum*

Clostridium botulinum Food Intoxication

Botulism is a food borne illness caused by the ingestion of food containing neurotoxin produced by *Clostridium botulinum*. It is a Gram positive rod shaped anaerobic spore forming soil bacterium which can grow in the pH range of 4.6 to 8.5. It produces highly potent exotoxin, a neurotoxin, and ingestion of a small quantity (few nanograms: 30-100 ng) can cause paralyses and death.

Toxin Types

Seven neurological toxin types (Type A to G) have been recognized based on toxin specificity. Of these types A, B, E and F are pathogenic to man. These are further divided into 2 types.

- Proteolytic types: Includes A, B, and F type toxin producers. These are heat resistant, NaCl tolerant, mesophilic and soil is the general habitat.
- Non-proteolytic types: Includes some strains of B, E, and F. these are heat sensitive, psychrotolerant, NaCl sensitive and aquatic environment is the natural habitat.
- Toxins produced by type A, B, E and F cause human botulism. Botulinal toxin is one of the most potent of all poisons and very low amounts (30 -100 ng) can cause death.

Clostridium botulinum Type E

C. botulinum Type E is associated with fish and fishery products and is primarily of marine origin. It is ubiquitous in natural environment including marine sediment as well as animals, birds and fish intestine. Toxin production by Cl. botulinum depends on the ability of cells to grow in food and autolyse to release toxins.

Factors Influencing Toxin Production

1. Toxin production depends on the factors influencing spore germination and growth. These include composition of food, moisture current, pH, O-R potential, salt content and temperature and storage conditions.
2. In heat processed foods *C. botulinum* are of great significance, as under-serilization can lead to survival of spore, which under suitable conditions germinate and make food toxic. Even seafood held at low temperature (refrigerated temperature) the type E can grow (at about 3°C). But fail to grow under frozen temperature.
3. For the intoxication to develop the preformed toxins should be consumed (food containing toxin). Food that has been eaten which was served without heating prior to serving can cause illness. The toxin is inactivated when heated at 60°C for 5 min. Botulism occurs rarely (incidence of botulism is not common) and most of the outbreaks are associated with fish (Type E).

Symptoms of Food Poisoning

Symptoms vary from mild illness to fatal condition within 24 hr, symptoms develop within 12-36 hr and include nausea and vomiting followed by neurological disorders like visual impairment, loss of normal function of mouth and throat, lack of muscle coordination, respiratory impairment leading the death. Onset of symptoms is rapid with type E botulism, while the severity is high with type A.

Treatment

Only known method of treatment is by administration of antitoxin and is effective only when administered before onset of symptoms.

Conditions Necessary for Outbreaks

1. Presence of spores of type A, B, E or F in food being canned or processed.
2. Food in which the spores can germinate and clostridia can grow and produce toxin.
3. Survival of spores because of inadequate heating in canning or inadequate processing.
4. Condition (environmental/storage) after processing that will permit germination of spores.
5. Insufficient cooking of food to inactive the toxin.
6. Ingestion of toxin bearing food.

Prevention of Outbreaks

1. Use of approved heat process for canned foods.
2. Rejection of all swollen or spoiled canned foods.
3. Refusal even to taste a doubtful food.
4. Avoidance of foods that have been cooked held but not well heated.
5. Boiling of suspected food for atleast 15 min to inactive toxin.

4.2. Non-indigenous Pathogens

1. *Salmonella*

Food poisoning caused by the ingestion of viable cells of genus *Salmonella* is called Salmonellosis. This is one of the frequently occurring bacterial infections. *Salmonella* are food/waterborne bacterial pathogens belonging to the family Enterobacteriaceae which cause clinical conditions ranging from mild gastrointestinal infections to systemic infections such as typhoid fever. *Salmonella* are food/waterborne bacterial pathogens belonging to the family Enterobacteriaceae which cause clinical conditions ranging from mild gastrointestinal infections to systemic infections such as typhoid fever. Over 4500 known serovars of *Salmonella* belong to either of the two species of *Salmonella, i.e. S. enterica* and *S. bongori*. Based on host adaptation, *Salmonella* serotypes are classified into three groups: (i) typhoidal (enteric) *Salmonella* responsible for typhoid and paratyphoid fever in humans and generally not pathogenic for animals; (ii) non-typhoidal *Salmonella* causing gastrointestinal illness in humans and a broad range of animals including other mammals, reptiles, birds and insects; and (iii) *Salmonella* causing infections in specific host such as *S. abortovis* and *S. gallinarum* to sheep and poultry respectively. Isolation of non typhoidal serovars from fish, shellfish and other seafood has been reported in several countries. Filter feeders such as oysters, shrimps and shellfish concentrate particulate matter including pathogenic bacteria present in the water which might explain the presence of *Salmonella* in seafood.

Characters

Salmonella are Gram negative, non-spore forming rods that ferment glucose usually producing gas. *Salmonella* spp. are facultatively anaerobic gram negative rod shaped bacteria belonging to the family, Enterobacteriacae. They are motile with perichous flagella. The *Salmonella pullorum* and *Salmonella guallinarum* are non motile strains. *Salmonella* are chemoorganotrophs that grow optimally at 37°C and catabolize D-Glucose with the production of acid and gas. They are oxidative negative, catalase positive, utilize citrate, produce H_2S, decarboxylate lysine and ornithine but do not hydrolyze urea.

Disease Symptoms

A large variety of foods including fish and shellfish are involved in salmonellosis. Food poisoning symptoms include nausea, vomiting, abdominal cramps, diarrhea, fever and headache. Symptoms occur 6~8 hrs after ingestion of contaminated food. Infective dose varies depending on serotype and ranges from few cells on 105 to cause illness. Persons of all age groups are susceptible. The infection is severe and prolonged among elderly and infants. The likelihood of infection depends on resistance of consumer, infectiveness of *Salmonella* strain and number of organisms ingested. More severe form of infection is typhoid and paratyphoid caused by *S. typhi* and *S. paratyphi. Septicemia* caused could be fatal.

Conditions Necessary for Outbreak

1. Food (fresh/processed) must contain or contaminated with *Salmonella*
2. Must be present in considerable numbers
3. Viable organisms must be ingested.

Prevention

- ☆ Avoidance of contamination of food from animals, humans, diseased food handlers and carriers.
- ☆ Destruction of the organism by heat
- ☆ Prevention of growth in foods by keeping at low temperature.
- ☆ Avoidance of consumption of warmed left over foods without refrigerator.

2. *Shigella*

Shigella is the bacterium that causes the disease shigellosis, also known as bacillary dysentery. *Shigella* is one of the most easily transmitted bacterial diarrheas, since it can occur after fewer than 100 bacteria are ingested. While reported cases of *Shigella* range between 14,000 and 20,000 annually. *Shigella sonnei* is the most common type of *Shigella*. It accounts for over two-thirds of cases of shigellosis in the United States.

Shigella bacteria are generally transmitted through a fecal-oral route. Foods that come into contact with human or animal waste can transmit *Shigella*. Thus, handling toddlers' diapers, eating vegetables from a field contaminated with sewage, or drinking pool water are all activities that can lead to shigellosis.

Symptoms of Shigella Food Poisoning

Symptoms of *Shigella* poisoning most commonly develop one to three days after exposure to *Shigella* bacteria, and usually go away within five to seven days. It is also possible to get *Shigella* but experience no symptoms, and still be contagious to others, a condition known as being asymptomatic.

Common Shigella Food Poisoning Symptoms

- ☆ Diarrhea: Diarrhea ranges from mild to severe. It is bloody in 25 to 50 per cent of cases and usually contains mucus,
- ☆ Fever

- Stomach cramps
- Rectal spasms

Complications from Shigella

Complications from shigellosis can include severe dehydration, seizures in small children, rectal bleeding, and invasion of the blood stream by the bacteria. Young children and the elderly are at the highest risk of death. The following is a list of specific complications caused by *Shigella*.

- Proctitis and Rectal Prolapse: The bacteria that causes shingellosis can also cause inflammation of the lining of the rectum or rectal prolapse.
- Reactive Arthritis: Approximately 3 per cent of patients with *Shigella* infection, most often those with *Shigella* flexneri, develop Reactive Arthritis. It occurs when the immune system attempts to combat *Shigella* but instead attacks the body. Symptoms of Reactive Arthritis include inflammation of the joints, eyes, or reproductive or urinary organs. On average, symptoms appear 18 days after infection.
- Toxic Megacolon: In this rare complication, the colon is paralyzed and unable to pass bowel movements or gas. Symptoms of Toxic Megacolon include abdominal pain and swelling, fever, weakness, and disorientation. If this complication goes untreated and the colon ruptures, the patient's condition can be life-threatening.
- Hemolytic Uremic Syndrome, or HUS: *Shigella* rarely results in HUS, which is more commonly a complication of Shiga toxin-producing *E. coli* infections. HUS can lead to kidney failure.

Diagnosis of Shigella

A *Shigella* infection is diagnosed through laboratory testing of a stool sample.

Shigella Food Poisoning Treatment

A *Shigella* infection usually goes away on its own in five to seven days, although bowel movements may continue to be abnormal for up to a month following infection. Antibiotics, however, can shorten the course of the illness. A doctor can prescribe antibiotics after testing a stool sample for the presence of *Shigella* bacteria.

Some strains of shigellosis are resistant to antibiotics, meaning that antibiotics might not always be an effective treatment. Antidiarrheal medication should be avoided, as it can actually make the illness worse.

Preventing a Shigella Infection

Frequent hand washing is key to preventing *Shigella*, since individuals can carry *Shigella* without noticing symptoms, and *Shigella* bacteria can remain active for weeks after illness.

Steps for Preventing the Spread of Shigella Infection

- ☆ If a child in diapers has shigellosis, wash your hands after changing their diaper and wipe down the changing area with disinfectant
- ☆ People with *Shigella* should not prepare food for others for at least two days after diarrhea has stopped
- ☆ Drink only treated or boiled water while travelling and only eat fruits you peel yourself
- ☆ Only swim in pools maintaining a chlorine level of 0.5 parts per million and stay clear of pools where children not yet toilet trained are swimming

3. *Staphylococcus aureus*

Staphalococcal food poisoning is caused by the ingestion of enterotoxin formed in food during growth of certain strains of *Staphylococcus aureus*. This bacterium produces an exotoxin called enterotoxin which causes gastroenteritis or inflammation of intestinal tract. *S. aureus* is a Gram positive cocci appearing as bunch of grapes. Most of the entertoxin producing strains are coagulase positive (coagulate blood plasma), produce thermostable nuclease, facultatively anaerobic and grow better aerobically than anaerobic condition. However, some coagulase negative strain also produces entertoxin. These produce six neurologically distinct entertoxin (A, B, C, C2 D and E) which differ in toxicity. Most food poising involves types A or D. Grow in the temperature range of 4–40°C with rapid growth taking placing in the temperature rage of 20-45°C.

Source of Staphylococcus to Food

Source of staphylococci to foods are mainly from human and domestic animals (primary reservoirs). Commonly encountered in skin, hair, nasal passage, wound and throat of humans. Food handlers are main source of food contamination with *S. aureus*. Besides, equipment and food contact surfaces also contribute for contaminatin with this organism. Generally, conditions in foods that favour the growth of *S. aureus* (at about 40°C) are more susceptible for food poisoning. Entertoxins are heat resistant and not inactivated at pasteurisation temperature. Normal cooking of food will not destroy toxin present in food. Seafoods support growth of *S. aureus* when contaminated. High salt tolerance and ability to grow at low water activity (0.86) helps this organism to have competitive advantage over other organisms especially in slat cured products such as smoked fish and salted fish.

Food Involved in Outbreaks

A variety of foods support the growth of *S. aureus*, and thus responsible for food poisoning outbreaks. The foods generally involved are Fish, meat, bakery products, salads *etc.*

Disease and Symptoms

Individuals differ in their susceptibility to food poisoning and incubation period is more than 4 hrs. Symptoms of food poisoning is characterized by salivation, nausea, vomiting, abdominal cramps, diarrhoea, blood and mucus in stools, headache *etc.* Mortality is very low and infected persons recover except in extreme cases.

Condition Necessary for Outbreak

- ✩ Food must contain entertoxin producing staphylococci
- ✩ Food must support the growth and toxin production by staphylococci
- ✩ Temperature must be favorable for growth, and there should be enough time lapses since contamination for toxin production.
- ✩ Entertoxin bearing food must be ingested.

Prevention of Food Poising Outbreaks

- ✩ Prevention of contamination of food with staphylococci
- ✩ Prevention of growth of staphylococci in foods.
- ✩ Avoiding consumption of contaminated food.

3. *Escherichia coli*

E. coli is the most common aerobic organism in the intestinal tract of humans and warm blooded animals. Generally the *E. coli* strains that colonize the gastrointestinal tract are harmless commensals, and they play an important role in maintaining intestinal physiology. However, certain strains of *E. coli* are pathogenic and cause gastrointestinal disturbances. Seafood is a major vehicle for transmission of several bacterial diseases. The faecal contamination of natural water bodies has emerged as a major challenge in developing and densely populated countries like India. Estuaries and coastal water bodies, which are the major sources of seafood in India, are often contaminated by the activities of adjoining populations and partially treated or untreated sewage from the townships is released into these water bodies. The fish harvested from such areas often contain human pathogenic microorganisms. In addition, poor sanitation in landing centers and the open fish markets exacerbates the situation. *Escherichia coli* has been traditionally recognized as an indicator organism of faecal contamination of water and seafood. Testing of seafood for the presence of *E. coli* is still a gold standard used to assess the faecal contamination in seafood processing plants in India and elsewhere. *E. coli* is a normal inhabitant of the intestinal tracts of all warm blooded animals. However, strains of human pathogenic *E. coli* have evolved that are recorded as causative agents of a broad range of human diseases compared to any other pathogenic bacteria. Currently, six categories of diarrheagenic *E. coli* have been reported: enterotoxigenic *E. coli*, enteropathogenic *E. coli*, enteroinvasive *E. coli*, enterohemorrhagic *E. coli*/Shiga toxin-producing *E. coli*, enteroaggregative *E. coli*, and diffusely adherent *E. coli*.

Classification of *E. coli*

E. coli can be serologically differentiated based on three major surface antigens as 'O' (somatic) antigen, 'H' (flagellar) antigen and 'K' (capsule) antigen. There are more than 700 serotypes of *E. coli*, identified based on O and H antigens, but not K antigen. The O antigen identifies the serogroup and H antigen identifies its serotype. Generally, a particular serogroup often fall into one category of pathogenic *E. coli*. But, serogroups of O55, O111, O126 and O128 are present in more than one category.

There are five categories of pathogenic *E. coli* as described below:

1. *Enteropathogenic E. coli (EPEC)*

EPEC are the first pathotype of *E. coli* that cause "watery diarrhea", but was originally termed as "Diarrheagenic *E. coli*". The major O serogroups are O55, O86, O111ab, O119, O125ac, O126, O127, O128 and O142. EPEC induce attaching and effacing (A/E) lesions in cells, to which they adhere and invade epithelial cells. Some types of EPEC are referred to as enteroadherent *E. coli*.

2. *Enterotoxigenic E. coli (ETEC)*

ETEC are the major cause of "infantile diarrhea" in regions with poor sanitation and frequently responsible for "traveler's diarrhea". ETEC colonize the proximal small intestine by fimbrial colonization factors (*e.g.* CFAI/CFAII) and produce LT or ST enterotoxin that elicits fluid accumulation and causes diarrhea. The major serotypes are O6, O8, O15, O20, O25, O27, O63, O78, O115, O128ac, O148, O159 and O167.

3. *Enteroinvasive E. coli (EIEC)*

EIEC are the major cause of "non bloody diarrhea and dysentery" similar to that caused by *Shigella* spp. The invasive capacity is associated with the presence of a large plasmid that encodes several outer membrane proteins. They invade and proliferate in epithelial cells of colon and cause widespread cell death. The major serogroups are O28ac, O29, O112, O124, O143, O144, O152, O164 and O167.

4. *Enteroaggregative E. coli (EAEC)*

EAEC are associated with "persistent diarrhea" in infants and children. They produce a characteristic pattern of aggregative adherence on the surface of HEp-2 cells as stacked bricks. The most common serotypes are O3, O15, O44, O77, O86, O92, O111, and O127. They produce a distinctive heat labile plasmid encoded toxin known as EAST (enteroaggregative ST) toxin. They also produce a hemolysin similar to those produced by EHEC.

In a large outbreak occurred in Germany in 2011 by the contaminated resulted in 50 fatalities and the causative agent for hemolytic uremic syndrome (HUS) was identified as *E. coli* O104:H4, which produced Stx2a and hence, grouped as "Shiga toxin producing *E. coli* (STEC)" strain (RobertKoch, 2011). It had 93 per cent genome homology with EAEC and also carried the aggR gene (transcriptional activator for expression of aggregative aggressive fimbriae - AAFI) found on an EAEC virulence plasmid.

5. Enterohemorrhagic E. coli (EHEC)

EHEC were first recognized as human pathogens in 1982, when *E. coli* O157:H7 was identified as the cause of two outbreaks of "hemorrhagic colitis". Among the different pathogenic *E. coli*, EHEC group is the most significant group that causes food borne illness. *E. coli* O157:H7 possess several characteristics uncommon to most other *E. coli* strains. They have inability to grow at >44.5°C, ferment sorbitol, produce β-glucuronidase, possession of a pathogenicity island known as the "locus of enterocyte effacement" (LEE), and carriage of a 60 MDa plasmid. All EHEC produce factors cytotoxic to African green monkey kidney (Vero) cell factors and hence named as "Verotoxins" or Stxs. *E. coli* capable of producing Stxs similar to the Stx produced by *Shigella* dysenteriae type I (O'Brien *et al.,* 1992) are sometimes named as Shiga toxin producing (STEC) strains. Among them, only those strains that cause hemorrhagic colitis are considered as EHEC. The major non-O157 serogroups are O26, O45, O103, O111, O121 and O145. They do not share the growth and metabolic characteristics of O157 EHEC, although they produce Stx, contain LEE and the large plasmid. Among the different pathogenic *E. coli*, EHEC group is the most significant group that causes food borne illness.

Disease Characteristics

E. coli O157:H7 infection includes non bloody diarrhea, hemorrhagic colitis (bloody diarrhea), and hemolytic uremic syndrome (HUS). Colonization occurs in the large bowel after an incubation period of 3-4 days following ingestion of the bacteria through food. Illness begins with severe abdominal cramps and non bloody diarrhea for 1-2 days, then progresses to bloody diarrhea that lasts for 4-10 days. Symptoms usually resolve after a week, but about 6 per cent of patients progress to HUS. One half of them require dialysis and 75 per cent require transfusions of erythrocytes and/or platelets. The fatality rate is about 1 per cent. HUS largely affects children leading to acute renal failure and the risk to develop HUS is about 15 per cent. Infectious dose is very low between 0.3 to 15 cfu of *E. coli* O157:H7 per gram. In an outbreak of *E. coli* O11:NM infection, which is a sorbitol fermenting EHEC, the count was less than one cell per gram.

Foodborne gastroenteritis caused by *E. coli* Prevention of food poison outbreaks

1. Good personal hygiene and health education of food handlers are essential in the control of disease.
2. Proper treatment (*e.g.* chlorination) of water and sanitary disposal of sewage.

3. Risk of infection can be minimized or eliminated by proper cooking before consumption.
4. The growth is generally inhibited in the presence of 4-5 per cent NaCl. Increased inhibition is seen at low temperature/or reduced pH.

4. *Campylobacter*

Foodborne gastroenteritis caused by Campylobacter Characters

- Campylobacter are represented by the species C. jejuni and C. coli are the major causative agents of bacterial gastroenteritis in humans worldwide.
- These are Gram negative curved rods, motile with single polar flagella at one or both ends.
- C. jejuni is associated with intestinal tract of warm-blooded animals and is often present in foods of animal origin through fecal contamination during processing.
- Foods of animal origin including milk and clams have been implicated in cases of food poisoning.
- This organism is unable to grow well and survive outside the animal host's environment, but ingestion of very low numbers can cause human illness.
- C. jejuni is microaerophilic a organism requiring low levels of oxygen for growth. Growth is inhibited at oxygen concentration of less than 3 per cent and more than 10-5 per cent, with best growth at 5 per cent. It grows within a narrow temperature of 30- 47°C with optimum temperature of 42-45°C, survives refrigerator temperature and is sensitive to freezing temperatures.
- Campylobacter are invasive and highly virulent requiring ingestion of only a few cells to cause illness.
- Though the actual mechanism of pathogenicity is not clearly understood, they are known to produce heat sensitive enterotoxins.

Symptoms of Illness

- Campylobacter infections occurs due to the consumption of foods of animal origin and contaminated milk, bakery products, clams and water.
- Symptoms of illness include abdominal pain, nausea, vomiting, cramps, diarrhoea, headache, fever and occasional bloody stools.
- The onset of illness varies from 1 to 10 days, and illness may persist for from 1 day to few weeks but most recover within a week.
- Campylobacter enteritis is self limiting in most cases and resolve within few days of onset of symptoms.

Prevention of Outbreak

- ☆ Proper cooking of food prior to consumption.
- ☆ Preventing cross contamination from raw meats to cooked foods.
- ☆ Chilling food rapidly.
- ☆ Following good sanitary measures while handling and processing of foods.

Chapter 5

Other Microbiological Hazards in Fish and Fishery Products

Biological Hazards

The presence of human pathogenic microorganisms in food has potential human health risk as the consumption of such food could cause food poisoning. Besides bacteria, presence of human pathogenic viruses and several other organisms such as parasites and worms or accumulation of toxic metabolites produced by toxic microalgae in fish and shellfish, and consumption such food could be cause of health related problems. These hazards of biological origin cause a serious health related problems. Hence, there is a need to free the food meant for human consumption form these biological hazards.

5.1. Mycotoxins

- ✰ Mycotoxins are toxic substances produced by a large number of molds (fungi), and are highly toxic to animals and potentially toxic to human beings. These are mutagenic, carsinogenic and some show toxicity restricted to specific organs.
- ✰ Mycotoxicosis is the syndrome resulting from the ingestion of mold toxin contaminated food. Of the several mold, members of Penicillium and Aspergillus sp produce mycotoxins as secondary metabolites during the end of exponential growth. Among the mycotoxins, aflatoxin is very important.

Aflatoxins

Aflatoxins are among the several mycotoxins produced by fungi and are more potent and most widely studied toxin. First reported in 1960 in England, when more than 1 lakh turkey poults died after feeding imported peanut meal from Africa. The causative agent was identified as *Aspengillus flavus* and the toxin was designated as aflatoxin. Aflatoxin is produced by certain stains of *A. flavus* and *A. parasiticus.*

Foods Supporting Aflatoxin Production

These fungi grow in a wide variety of foods/substances – dairy products, bakery products, fruit juices, cereals, forage crops, dry fish *etc.* More common in peanuts, cotton seeds and corn as fungal invasion, growth and mycotoxin production takes place before harvesting. Generally, contamination and aflatoxin production is related to insect damage, humidity, weather condition and agriculture practices. These produce aflatoxin in mesophilic temperature range (25-40° C) and water activity of 0.85.

Aflatoxin Types

Aflotoxin is composed of 4 major toxic substances (toxic metabolites responsible for aflatoxin) namely B1, G1 and B2 and G2 based on whether they fluoresce blue (B) or green (G) under UV light. *A. flavus* produces B1 and B2 (AFB1and AFB2) while, *A. parasiticus* produces all four aflatoxins (B1, B2, G1, G2). AFBI is most potent of all aflatoxins. Fishes fed aflatoxin containing feed (contaminated fish meal and other feed ingredients, contaminated feed) affects growth and also carcinogenic in trout. International guidelines permit maximum of of 30 ppb of aflatoxin in foods and feed ingredients.

5.2. Parasites

Human Parasitic Worms in Seafood

Parasites are involved in substantial numbers of seafood related illness especially in countries where seafood is eaten raw. More than 40 million people are affected worldwide every year by trematode infections. Parasite worms are commonly associated with seafood and in certain countries of Eastern Asia some parasites are endemic. The parasites of concern belong to different groups, nematodes, cestodes and trematodes. They have complex life cycles with a definite host (man or animals) and a primary intermediate host (generally snails) and a secondary intermediate host (generally fish or crustaceans). The sexual maturity and the release of eggs take place in man or animals (the definite host) while the larval stages are reached and completed in intermediate hosts. The larval stages in second intermediate host (fish or crustaceans) infect humans when fish or shellfish harbouring them are consumed raw. The larval stages in second intermediate host are known as cercariae or metacercariae. Round worms that include *Anisakis simplex* and *Pseudoterranova decipiens* are the most important species of anisakid worms infecting humans worldwide while the rest are restricted to some geographical regions only. The symptoms include sudden onset of abdominal pain, nausea and vomiting. Anisakid worms are also a major problem in Europe and the US. The infective larval stage is found in the intestine and the muscle of the fish. Therefore rapid chilling following catch and degutting will substantially reduce the worm burden. The fish intended to be eaten raw must be frozen at -20°C for at least 24 h. Salting or marinating will destroy the worms.

Among tapeworms transmitted through seafood, *Diphyllobothrium* is the most important group, *D. latum* and *D. dendriticum* being are the most common species infecting humans. *D. latum* can grow to a length of more than 10 m and competes for vitamin B12 in human host causing pernicious anaemia. Cases have been reported from US and Japan. The worms are found in the small intestines of fish. The symptoms of infections include abdominal pain, vomiting and diarrhoea. However, the patient may, in several cases, remain asymptomatic for many years.

The trematode worms are most important and the largest group of parasites infecting humans. It has been estimated that about 40 million people are infected annually with these parasites. The infections occur mainly in countries where the parasites are endemic and present in large numbers in fish harvesting areas. The parasites occur in the skin of infected fish or shellfish (molluscs or crabs), as metacercariae. The trematode worms are classified based on their organ tropism as liver flukes (*Clonorchis sinensis*, and *Opisthorchis viverrini*, *O. felineus*), lung flukes (*Paragonimus* sp.) and blood flukes (*Heterophyes* and *Metagonimus* spp.) The liver flukes grow and reproduce in the liver and the eggs are passed out in the faeces of infected individual. The most common human lung flukes *Paragonimus westermani* and *P. skrjabini* are endemic in countries such as Thailand, Japan, China and Korea. Snails are the first intermediate host while crustaceans such as freshwater crabs serve as second intermediate host. The infections are acquired by eating crabs raw and the larvae enter into the body cavity after passing trough intestinal epithelium and get lodged into the pulmonary cavity, where they grow into adult worms in the lungs. The intestinal flukes of importance commonly encountered in seafood are *Metagonimus yokagawai* and *Heterophyes heterophiles*, though more than 50 species are known to infect human. Though both the worms have snails as first intermediate host in their life cycle, they differ in their final host. For *Metagonimus* sp., freshwater fish are the second intermediate host while for *Heterophyes* sp., brackish water fish serve as the second intermediate host. The traditional habits of eating raw fish make control of infections difficult. The use of human or animal faeces for manuring fishponds maintain constant source of infection.

Foodborne Parasites

Several human parasites are associated with fish and shellfish and consumption of such foods leads to transmission of parasites to humans. These include members belonging to the group protozoans, flat worms and round worms. These parasites are easily detectable in food because of their larger size than bacteria, unable to proliferate in food, not possible to grow on culture media, and complete life cycle by passing through one or more animal hosts. Consumption of foods contain these parasites could lead to health risks.

5.2.1. Protozoan Parasites

The important protozoan parasites involved in human illness are members of the genus *Giardia, Entamoeba, Toxoplasma, Sarcocystis* and *Cryptosporidium.*

Giardia sp.

The flagellated protozoan *Giardia lamblia* exists naturally in aquatic environment and causes human illness called giardiasis.

Characters

1. The Giardia cells (trophozoites) produce cysts which infect humans through water and food. T rophozoites are characterized by the presence of eight flagella arising on the ventral surface near the paired nuclei.
2. The cysts are pear shaped, with the size ranging from 8-20 µm in length and 5-12 µm in width.
3. The excystation of ingested cysts in upper small intestine releases the trophozoites which invade intestinal wall and bile duct.
4. Growth is generally favoured by high levels of bile juice in the duodenum and upper jejunum.

Symptoms of Illness

- ☆ Giardiasis is highly contagiouis.
- ☆ The disease symptoms appear after a incubation period of 6-13 days, cysts appear in stools in 3-4 weeks, and symptoms may last for months to a year or more.
- ☆ The leading symptoms of infection are diarrhea, fatigue, abdominal cramps, fever, nausea, vomiting and weight loss.

Control Measures

1. The minimum dose required for infection in humans is 10 cysts or less.
2. Avoiding sewage contamination of natural waters.
3. Avoiding consumption of sewage contaminated water and fish harvested from sewage polluted waters.

Entamoeba histolytica

This protozoan parasite is responsible for amebiasis or amoebic dysentery and is transmitted mainly through fecal-oral route, and also through water, food handlers and foods.

Characters

1. *Entamoeba histolytica* is an motile aerotolerant anaerobe, trophozoites are of 10-60 µm in size and lack mitochondria, cysts non motile and are of 10-20 µm in size.
2. Trophozoites do not persist in the environment but cysyts can survive in sewage sludge for up to 3 months and transmitted through water and food.
3. The trophozoites in intestine adhere to host cell glycoprotein, cause abscesses in intestinal mucosal cells and ulcers in colon, multiply by binary fission in large intestine, encysts in ileum, and produce enterotoxic protein.

Disease Symptoms

1. After ingestion of the parasite.
2. The symptoms of disease occur after 2-4 weeks of incubation of ingested parasite and symptoms persist for several months.
3. The common symptoms include mucus and blood in stool, abdominal pain, fever, severe diarrhea, and vomiting and weight loss.
4. Affected persons can be treated using amebicidal drugs such as metronidazole and chloroquine.

Toxoplasma gondii

The obligate intracellural protozoan, *Toxoplasma gondii*, is responsible for causing a disease calld toxoplasmosis.

Characters

1. Domestic and wild cats are the definitive hosts for this parasite and serve as primary source of human infection.
2. Oocysts are transmitted from cat to cat and infect all other animals, and survive over a year in warm, moist environment.
3. Clinical symptoms in humans are caused by as few as 100 oocytes. Encysted form can remain latent in humans.

Disease Symptoms

1. Toxoplasmosis is regarded as universal infection and symptomless in most individuals.
2. When severe, symptoms of fever with rash, headache, muscle ache and pain and swelling of lymph nodes is noticed.

Control Measures

- ✰ Avoiding environmental contamination with cat feces.

- ☆ Avoiding consumption of meat and meat products containing viable tissue cysts.
- ☆ Heating food above 60°C which destroys cysts.

***Sarcocystis* sp.**

1. *Sarcocystis* sp. cause sarcocystosis in humans.
2. Cattle and pigs serve as intermediate host while humans are definitive hosts.
3. Species involved in disease include *Sarcocystis hominis* which is associated with cattle, and Sarcocystis suihomonis, associated with pigs.
4. Infection occurs through consumption of infected porcine and bovin meat, contaminated food and water.
5. Sarcocysts in humans transform to bradyzoites which penetrate small intestine, reproduce sexually resulting in sporocysts and pass out through feces.
6. The sporocysts when ingested by pigs or bovines release sporozoites and spread throughout the body.
7. These multiply asexually in skeletal and cardiac muscle producing sarcocysts.
8. Sarcocysts are visible to naked eye as they reach a size of about 1cm.

Disease Symptoms

1. Symptoms occur within 3-6 hr of infection and consist of nausea, stomachache, and diarrhea.

5.2.2. Pathogenic Flatworms and Round Worms

Occurrence of flatworms and round worms is common in fish and shellfish and over 50 species of helminth parasites from fish and shellfish are known to cause disease in human beings. These parasitic helminths have complicated life cycles during their development involving a number of intermediate hosts. Generally, sea snails or crustaceans serve as first intermediate host, marine fish as second intermediate host, and mammals as the final host for sexually mature parasites. In between these hosts, one or more free living stages may occur. Infection of human may be part of this life cycle or it may be a side track causing disruption of the life cycle.

The pathogenic flatworms and round worms transmitted by fish and shellfish are:

Nemotodes or Roundworms	*Fish/Shellfish Involved*
Anisakis simplex	Herring
Pseudoterranova dicipiens	Cod

Gnathostoma sp.	Fresh water fish, frog
Capillaria sp.	Freshwater fish
Angiostrongylus sp.	Freshwater prawns, snails, fish

Cestodes or Tapeworms

Diphyllobothrium latum	Freshwater fish
D. pacificum	Marine fish

Trematodes or Flukes

Clonorchis sp.	Freshwater fish, snails
Opisthorchis sp.	Freshwater fish Paragonimus sp. Snails, Crustaceans, Fishes
Echinostoma sp.	Clams, Freshwater fishes, Snails
Heterophyes sp.	Snails, Freshwater fish, Brackishwater fish
Metagonimus yokagawai	Freshwater fish

A. Nematodes or Round Worms

1. Round worms or nematodes are common and found in marine fish all over the world.
2. Anisakis simplex is commonly known as "herring worm" and *Pseudoterranova dicipiens* is known as "Cod worm".
3. Live worms when ingested by humans penetrate into the wall of the gastrointestinal tract and cause an acute inflammation ("herring worm disease").
4. *Gnatwhostoma* sp. is another nematode found in freshwater fish in Asia. The ingested larvae migrate from the stomach to other body regions (subcutaneous sites in the thorax, arms, head and neck) and induce a creeping sensation and edema.
5. *Capillaria philippinensis* infection in humans causes severe diarrhea and possible death due to fluid loss.
6. *Angiostrongylus cantonensis* is an common nematode in Asia and is associated with freshwater fish, snails and prawns, and known to cause meningitis in humans.

B. Cestodes

1. Very few cestodes or tapeworms are transmitted to humans through fish.
2. The broad fish tapeworm, *Diphyllobothrium latum* is a common human parasite affecting the intestinal tract of humans.
3. The related species (*D. pacificum*) is transmitted by marine fish.

4. These parasites are transmitted through the consumption of raw or semi processed fish.

C. Trematodes

- ☆ Some of the trematodes or flukes are common in Asia. The liver fluke (*Clonorchis sinensis*) is a common parasite infecting bile-ducts in the liver of humas in Asia.
- ☆ Flukes such as *Metagonimus yokogawai* and *Heterophyes heterophies* infect the intestines of the final host (humans) causing inflammation, symptoms of diarrhea and abdominal pain. The adult oriental lung fluke, *Paragonimus* sp. lives in cysts in the lungs of final hosts.

Control of Parasitic Infection

All infections involving parasites are transmitted by eating raw or uncooked fish products. Control measures to reduce the public health problem related to the presence of parasites include:

1. Avoidance of capture and consumption of nematode-infected fish by selecting specific fishing grounds, specific species or specific age groups.
2. Sorting and removal of nematode-infected fish or removal of nematodes from fish.
3. Application of techniques to kill nematodes in the fish flesh (Ex. Freezing).

5.3. Viruses

The incidence of foodborne outbreaks of viral gastroenteritis is quite common. Viral disease transmission to human beings via consumption of seafood has been known since the 1950s and human enteric viruses are implicated as a major cause of shellfish- associated disease conditions. Presently, more than 100 known enteric viruses are excreted along with faeces by infected individuals and finally find their way in to domestic sewage. The seafood associated viral infection causing illness are ;

- ☆ Hepatitis- type A (HAV)
- ☆ Norwalk virus (small, round structured)
- ☆ Snow Mountain agent
- ☆ Calcivirus
- ☆ Astrovirus
- ☆ Non-A and Non-B viruses

Survival of Viruses in the Environment and in Food

The survival of viruses in the environment and in food is dependent on factors such as temperature, salinity, solar radiation, and presence of organic solids. Enteric viruses survive much longer than coliform bacteria, and for several months in seawater at temperatures <10°C. Thus, there is little or no correlation between presence of virus and coliforms which are the common indicator bacteria for faecal pollution. All enteric viruses are resistant to acid pH, proteolytic enzymes and bile salts in the gut.

Hepatitis type A virus is one of the more heat stable viruses and has an inactivation time of 10 min. at 60°C, thus are able to survive some commonly used culinary preparation methods (steaming, frying).

Enteric viruses are resistant to some commonly used disinfectants (*e.g.* phenolics, ethanol, quaternary ammonium compounds), but sensitive to halogens (*e.g.* chlorine, iodine). Ozone is highly effective in clean water.

Foods Involved in Viral Infection

Seafood-associated viral infections are mainly due to the consumption of raw or improperly cooked molluscan shellfish. HAV transmission has been attributed to unsanitary practices during processing, distribution or food handling. One of the largest outbreaks of food-borne illnesses involving 2, 90,000 cases was reported in China in 1988 due to the consumption of contaminated and inadequately cooked clams.

Virus Associated with Outbreaks in Shellfish

Virus	*Shellfish*
Hepatitis A Virus (HAV)	Clam, oysters, mussel
Norwalk virus	Oysters
Snow round virus (Astrovirus, Calcivirus, Parvovirus, SRSV)	Oysters, cockles, mussel
Small round structured virus (SRSV)	Oysters

Viruses Associated with Seafoods

Viruses constitute an important cause of seafood borne diseases. The intensity of human illness is known only from developed countries. The major limitation in diagnosing viral infections is the nonavailability of techniques or facilities for detecting or identifying viral pathogens. The recent advance in molecular biology has helped to develop or improve the detection techniques for viruses but these are still not used commonly in developing countries. Viruses being obligate intracellular parasites cannot multiply in seafood, but molluscan shellfish concentrate these infectious agents in their tissues. This is the reason why molluscan shellfish are associated with most of the well-recorded outbreaks. Nevertheless, very low numbers (1-100) of viruses are known to be capable of causing infections in humans making these organisms all the more important as agents of human illness. The transmission is through fecal-oral route with infected individuals becoming source of viruses to the aquatic environment. The viruses causing human infections are classified into two groups, viral gastroenteritis and viral hepatitis. The illnesses due to viruses are mainly prevalent in countries where eating raw shellfish is a dietary habit. Though there are more than 110 different viruses known to be excreted in human feces, collectively called the "enteric viruses", hepatitis A virus and Norwalk viruses are most commonly involved in seafood related illnesses followed by rotaviruses and astroviruses which cause infections in children.

In recent years, viruses have been increasingly recognized as important causes of foodborne disease. One category of implicated foods is those that are minimally processed, such as bivalve molluscs and fresh produce. In addition, many of the documented outbreaks of foodborne viral illness have been linked to contamination of prepared, ready-to-eat food by an infected food handler. These are typically contaminated with viruses in the primary production environment. The viruses implicated in food-borne disease are the enteric viruses, which are found in the human gut, excreted in human feces, and transmitted by the fecal-oral route. Many viruses are found in the gut, but not all are recognized as food-borne pathogens. The enteric viral pathogens found in human feces include noroviruses (previously known as Norwalk-like viruses), enteroviruses, adenoviruses, hepatitis A virus (HAV), hepatitis E virus (HEV), rotaviruses and astroviruses, most of which have been associated with food-borne disease outbreaks. Noroviruses are the major group identified in food-borne outbreaks of gastroenteritis, but other human derived and possibly animal-derived viruses can also be transmitted via food.

Enteric Viruses

The human enteroviruses, classified within the Picornaviridae family, have been shown to be present in human feces and in domestic sewage. Approximately 66 immunologically distinct serotypes of the human enteroviruses are known to cause infections in humans, including the polioviruses, group A and B coxsackie viruses, and the echoviruses, and the more recently designated enterovirus serotypes 68 to 71. Many of the infections caused by human enteroviruses are asymptomatic, although, when symptomatic, a wide range of clinical syndromes, including fever, paralysis, meningitis, poliomyelitis, respiratory disease, and diarrhea, have been reported. Numerous studies have reported the detection of human enteroviruses in molluskan shellfish from estuarine areas, both open and closed to harvesting, and they are, by far, the most commonly isolated viral agent in shellfish. While other members of the enterovirus group have been reported in association with food and water, only a small number of foodborne disease outbreaks caused by coxsackie and echoviruses have been recorded. Unfortunately, there is no correlation between the presence of human enteric viruses and the level of fecal coliforms, the common index of sanitary quality in shellfish or their harvesting waters.

Hepatitis A Virus (HAV)

Several viruses cause hepatitis but only two, HAV and HEV, are transmitted by the fecal-oral route and are listed as "Severe Hazards" in U.S. Food and Drug Administration's Food Code. The incidence of HAV infection varies considerably among and within countries. In most developing countries, where hepatitis A infection is endemic, the majority of persons are infected in early childhood, when the infection is generally asymptomatic. Virtually all adults are immune. In developed countries, however, HAV infections are less common as a result of improved standards of living. Very few persons are infected in early childhood, and the majority of adults remain susceptible to infection by HAV. Later in life, HAV infection may result in a more severe disease outcome. As a result, the potential risk of outbreaks of HAV is increased in these regions. The hepatitis viruses are so named because they infect the liver, rather than sharing phylogenetic or morphological similarities, and each of the five different hepatitis viruses is classified in a distinct viral family. HAV causes hepatitis A, a severe food and waterborne disease that was formerly known as infectious hepatitis or jaundice. The virus is primarily transmitted by the fecal-oral route, but can also be transmitted by person-to-person contact. Hepatitis A infection occurs worldwide and is especially common in developing countries, where more than 90 per cent of children have been reported to be infected by 6 years of age.

HAV is a 27- to 32-nm, nonenveloped, positive-sense, single-stranded RNA virus with a 7.5 kb genome, icosahedral capsid symmetry, and a buoyant density in cesium chloride of 1.33–1.34 g/ml. The virus is classified in the Picornaviridae family in its own distinct genus, Hepatovirus. But, it was formerly classified in the Enterovirus genus as Enterovirus. It has a structure similar to that of other picornaviruses. There is one species, HAV, with two strains or biotypes: human HAV and simian HAV. These two distinct strains are phylogenetically distinct and have different preferred hosts. Human HAV infects all species of primates including humans, chimpanzees, owl monkeys, and marmosets, whereas simian HAV infects green monkeys and cynomolgus monkeys.

Hepatitis E Virus (HEV)

HEV is believed to be a major etiologic agent of enterically transmitted non- A, non-B hepatitis in humans worldwide. The virus is transmitted by the fecal-oral route and occurs widely in Asia, northern Africa, and Latin America, including Mexico, where waterborne outbreaks are common. It is a 32- to 34-nm, nonenveloped, positive-sense, single-stranded RNA virus with a linear genome of 7.2kb. The capsid symmetry is icosahedral, and the buoyant density in potassium tartrate–glycerol gradient is 1.29 g/ml. HEV was originally classified in the Caliciviridae because of similarities in structural morphology and genome organization. The virus was then reclassified in the family Togaviridae because of similarities between the replicative enzymes of HEV and the togaviruses. However, the current International Committee on Taxonomy of Viruses (ICTV) classification places HEV under a new family, Hepeviridae, genus Hepevirus. The major mode of transmission appears to be contaminated water. Secondary person-to-person transmission has been estimated at 0.7–8.0 per cent and is relatively uncommon. Food-borne outbreaks of HEV are the most common in developing countries with inadequate environmental sanitation. Large waterborne outbreaks have also been reported in Asian countries. The waterborne transmission route has been proved, and there have been reports of possible foodborne outbreaks in China.

Norovirus and Sapovirus

Noroviruses, previously known as small round structured viruses (SRSVs) and Norwalk-like viruses (NLVs), are now the most widely recognized viral agents associated with food-borne and waterborne outbreaks of nonbacterial gastroenteritis and probably the most common cause of foodborne disease worldwide. These viruses cause epidemic viral gastroenteritis resulting in large outbreaks. Noroviruses are primarily transmitted through the fecal-oral route, by consumption of fecally contaminated food or water, or by direct person-to person spread. Secondary spread may also occur by airborne transmission. Outbreaks commonly occur in closed community situations such as rest homes, schools, camps, hospitals, resorts, and cruise ships and where the food and water sources are shared. It has been estimated that noroviruses are responsible for approximately 60 per cent of food-borne disease in the United States, including more than 9 million cases, 33 per cent of hospitalizations, and 7 per cent of deaths related to foodborne disease each year.

Food-borne norovirus outbreaks resulting from preharvest contamination of foods such as shellfish and postharvest contamination through food handling have been reported worldwide. Presymptomatic infection in food handlers has been shown to cause outbreaks of food-borne norovirus infection. Determination of the original source of the virus is often problematic because several modes of transmission frequently operate during norovirus gastroenteritis outbreaks. Although the initial transmission route may be through consumption of contaminated foods, secondary transmission via direct contamination of the environment or person-to-person contact also often occurs. Studies using human volunteers showed that norovirus retains infectivity when heated to 60°C for 30 min and therefore is not inactivated by pasteurization treatment. The virus also retains infectivity after exposure to pH 2.7 for 3 h at room temperature. Further evidence of its resistance to low pH was shown when norovirus was exposed to heat treatment and subsequent marination at pH 3.75 in mussels for 1 month. There are anecdotal reports of people developing gastroenteritis after eating pickled shellfish. Norovirus, like other enteric viruses, remains infectious under refrigeration and freezing conditions, appears to survive well in the environment, and is resistant to drying. These viruses are also resistant to treatment with 3.75 to 6.25 mg chlorine/L, which is equivalent to free residual chlorine of 0.5 to 1.0 mg/ml, a level of free chlorine consistent with that generally present in a chlorinated drinking water supply. However, the viruses were inactivate after treatment with 10 mg/L of chlorine, which is the concentration applied to water supplies after a contamination event.

The sapoviruses, formerly described as the "Sapporo-like viruses," or SLVs, also belong to the Caliciviridae family and cause gastroenteritis among both children and adults, although association with food-borne transmission is rare.

Rotavirus

Rotaviruses are the major cause of severe diarrhea and gastroenteritis in infants and young children. It is estimated that rotaviruses cause more than 130 million cases of diarrhea in children under 5 years of age annually worldwide. Rotaviral infection is a particularly serious problem in developing countries where up to 6,00,000 deaths occur annually among children. Rotaviruses are transmitted by the fecal-oral route and cause disease in both humans and animals, especially domestic animals, with subsequent serious economic loss. Although the animal and human strains are usually distinct, some human strains are closely related to animal strains, and cross species infections do occur. Rotaviruses are classified in the genus Rotavirus in the family Reoviridae, a large family composed of nine genera. Electron micrographs of rotaviruses show a characteristic wheel-like appearance, hence the name rotavirus, derived from the Latin meaning "wheel". Rotaviruses are 60 to 80 nm, nonenveloped, linear segmented doublestranded RNA viruses with icosahedral, capsid symmetry. The 16 to 27 kb, genome is enclosed by a triple-layered capsid composed of a double protein shell and an inner core. Eleven segments of DNA code for six structural and five nonstructural proteins. Two of the structural proteins, VP7 (glycoprotein) and VP4 (protease or P protein), comprise the outer shell of the capsid and are important in virus infectivity.

Astrovirus

Astroviruses are distributed worldwide and have been isolated from birds, cats, dogs, pigs, sheep, cows, and human. The main feature of astrovirus infection in both humans and animals is a self-limiting gastroenteritis. The astroviruses are a common cause of human gastroenteritis, with most cases of infection detected in young children under 1 year of age. The astroviruses were first recognized in 1975 and were named according to their star-like appearance under the electron microscope. They belong to the family Astroviridae, and human astrovirus is the single type species in the genus Mamastrovirus. Astroviruses are 28 to 30 nm, spherical, nonenveloped, positive-sense, and single-stranded RNA viruses. Epidemiological evidence of transmission by foods is limited, but infections via contaminated shellfish and water have been reported. In 1991, a large outbreak of acute gastroenteritis occurred in Japan involving thousands of children and adults from 14 different schools.The outbreak was traced to food prepared by a common supplier for school lunches. There are several Japanese reports of astrovirus genomes identified in shellfish, and there is evidence that astroviruses appear to contribute to food borne outbreaks of gastroenteritis mainly through the consumption of contaminated oysters.

Astroviruses are resistant to extreme environmental conditions. Their heat tolerance allows them to survive 50°C for 1h. At 60°C, the virus titer falls by 3 $\log_{10}$ and 6 $\log_{10}$ after 5 and 15 min, respectively. The virus is also stable at pH 3.0 and is resistant to chemicals, including chloroform, lipid solvents, and alcohols and to nonionic, anionic, and zwitterionic detergents.

Adenovirus

The adenoviruses are widespread in nature infecting birds and mammals including human. They commonly cause respiratory disease but may also be involved in other illnesses such as gastroenteritis and conjunctivitis. In particular, the enteric adenoviruses cause gastroenteritis and are the second most important cause, after rotaviruses, of acute gastroenteritis in children under 4 years of age. Adenoviruses can be transmitted from person-to-person by direct contact or via fecal-oral, respiratory or environmental routes. Adenoviruses belong to the Adenoviridae family and are classified into two genera: the Mastadenovirus, which infects mammals, and the Aviadenovirus, which infects birds. Adenoviruses are 80 to 110 nm, nonenveloped, linear double-stranded DNA viruses with icosahedral symmetry and a genome of 28–45kb.

Adenoviruses have been identified in a variety of environmental samples, including wastewater, sludge, shellfish, and in marine, surface and drinking waters. No food-borne or waterborne outbreaks associated with the enteric adenoviruses have been reported, but, as these viruses are common in the environment, it is possible that disease has occurred but the source of infection has not been recognized. There is no documented evidence for food-borne transmission or disease resulting from consumption of adenovirus-contaminated shellfish. Adenoviruses are resistant to various chemical and physical agents including lipid solvents and to adverse pH conditions. They can withstand freeze-thawing several times without a significant decrease in titer but are inactivated after heating at 56°C for more than 10 min. The adenoviruses are capable of prolonged survival in the environment and are considered to be more stable than enteroviruses in many environmental situations.

Enteroviruses

Enteroviruses include polioviruses, coxsackie A and B viruses, and echoviruses, many of which are culturable. They are transmitted by the fecal oral route and are excreted in feces but do not generally cause gastroenteritis. Polioviruses were the first viruses to be shown to be food-borne, but because of the mass immunization campaigns, virulent wild-type strains are now rarely seen. Outbreaks of food-borne illness associated with coxsackieviruses and echoviruses have been reported. The enteroviruses are 28 to 30 nm, nonenveloped, positive-sense, singlestranded RNA viruses with icosahedral symmetry and a genome of 7.2– 8.4 kb. They are classified in the large Picornaviridae family, and seven species have been designated within the Enterovirus genus, namely bovine enterovirus, human enterovirus A, human enterovirus B, human enterovirus C, human enterovirus D, poliovirus, porcine enterovirus A, and porcine enterovirus B.

The first recorded outbreak associated with food-borne viruses was an outbreak of poliomyelitis linked to consumption of raw milk in 1914. The widespread introduction of pasteurized milk in the 1950s decreased transmission by this route. There have been very few recorded food-borne outbreaks associated with enterovirus infection despite the regular occurrence of enteroviruses in the environment. Enteroviruses, including echoviruses and coxsackie A and B viruses, have been isolated from sewage, raw and digested sludge, marine and fresh waters, and shellfish.

The enteroviruses are resistant to environmental stressors including heat, adverse pH, and chemicals. Because they are easily cultured in vitro and are stable in the environment, live attenuated vaccine strains of poliovirus have been used as indicator viruses for the presence of other virulent enteric viruses in food and water. They have also been used extensively in environmental and food virology research for methods development and to gather information on virus recovery, persistence, and behavior in these settings.

Conclusion

Foodborne diseases are a major problem worldwide. In Japan, clams are usually boiled before they are consumed in soups. However, boiling to open the clam may not inactivate the virus; in addition, some areas in Japan do not boil clams before eating them. Despite the fact that viruses are the most common pathogens transmitted via food. There is no systematic inspection and legislation exist that would set up virological criteria for food safety, regarding the presence of viruses in the food chain. Accordingly, the education of food industry managers, producers, distributors, and consumers about hygienic regulations and conditions of food production and processing (the use of non-infected water for watering and food processing, clean utensils *etc.*) and particularly their compliance are essential.

5.4. Marine Toxins

Marine toxins are natural toxins found in fish and shellfish. Consumption of toxin containing seafood results in various kinds of seafood borne illness. Marine toxins or biotoxins are produced by naturally occurring marine algae/ phytoplankton and marine organisms become toxic by feeding on these toxic phytoplankton.

Shellfish and finfish toxins continue to be a threat to public health and local fisheries around the world. In the last decade seafood monitoring has intensified and detection methods have continued to develop in selectivity and sensitivity. While some of these toxins only occur in seafood from a restricted geographical region, the export of exotic seafood that may contain such toxins can lead to unexpected public health issues. Increasingly sophisticated instrumentation (*e.g.* Liquid Chromatography/Mass Spectrometry/Mass Spectrometry -LC/MS/MS) is employed to identify and confirm the presence of known toxins, and is often necessary in light of the complex toxin profiles that can occur in shellfish tissue. Another important development is the introduction of sensitive and selective antibody-based detection methods for a number of toxins. Although most of these antibody-based test kits have yet to receive international regulatory approval, several scientific reports have described their efficacy and potential. An emerging technology is the use of immunosensors and some practical applications for detecting shellfish toxins have been reported. It remains to be seen if this approach gains momentum in the future. Nevertheless, more traditional methods such as the mouse bioassay continue to be used in many countries to warn of the presence of toxins, known and unknown.

5.4.1. Paralytic Shellfish Poisoning (PSP) Toxins

The PSP toxins are arguably the most recognised group of marine toxins and continue to be regarded among the most dangerous, with intoxications occurring in many countries even today. Saxitoxin (STX) and neosaxitoxin (neoSTX) are the parent members of the group and are the most toxic. Around twenty PSP toxins are known, all of which contain a tetrahydropurine skeleton. Not all the toxins found in shellfish extracts are microalgal products; some are the result of biotransformation processes by the filter-feeding host. PSPs are water-soluble and heat stable so cooking has little or no effect on PSP toxicity. The N-carbamoyl sulfates are readily hydrolysed to STX, neoSTX and Gonyautoxins (GTX). All PSPs degrade rapidly in alkaline conditions.

Origins and Distribution

The PSP toxins are produced by several dinoflagellate species belonging to the genera Alexandrium, Gymnodinium catenatum, and a tropical species Pyrodinium bahamense. Many of these species are extremely common and are foundthroughout the world in a variety of cold, temperate, and sub-tropical waters and habitats. Consequently, a great variety of shellfish from all latitudes, even crustaceans such as lobsters, can become contaminated with these toxins.

Toxicology

PSP toxins affect the central nervous system of mammals, and humans seem to be particularly susceptible. PSPs bind to site 1 of sodium channels reversibly blocking sodium conductance in nerve and muscle membranes by binding to a specific receptor site located on the outside surface of the membrane close to the ion channel opening. Neurophysiological studies have shown that PSP toxins all possess the same mode of action though some derivatives such as STX and neoSTX are considerably more potent than the C-toxins. However, the less potent C toxins are readily hydrolysed by acid to the more potent GTX toxins. Human symptoms begin with tingling and numbness of the lips, tongue, and fingers within 5 - 30 minutes of consumption, leading to paralysis within a few hours and in extreme cases death by asphyxiation due to respiratory paralysis. Cooking does not destroy these compounds and although some of the toxins may be extracted into the cooking water, this is never sufficient to render the seafood safe to consume.

Detection Methods

The safe limit set by most countries for PSP toxins is 800 μg/kg (0.8 μg/g) of shellfish or seafood tissue. The mouse bioassay has been in place for decades and the method has been standardised, and while simple and relatively inexpensive to operate, the detection limit (4 μg/g) of the method is close to the safe level and does not provide any information on the profile of PSP toxins present in the extract. However it remains as the monitoring method of choice in many countries. There continues to be a significant effort applied to the development of alternative chemical and biochemical detection methods for these toxins. One approach has been to take advantage of the ready conversion of these toxins to fluorescent derivatives. Two approaches have been taken here, namely oxidative conversion to fluorescent derivatives after High Performance Liquid Chromatography (HPLC) analysis (post-column derivatisation), or conversion before HPLC separation (pre-column derivatisation) and these have recently been thoroughly reviewed. Early post-column methods were somewhat time-consuming and required three separate HPLC runs to complete the analysis. Alternative HPLC methods have subsequently reduced this to a single HPLC run. After some development and various trials, a pre-column method has received Association of Analytical Communities (AOAC) approval, and has also been evaluated by a number of EU laboratories. While the chemical detection methods are most promising for the future, a criticism has been that such methods are time consuming, and more standards of individual PSP toxins are required to validate the complete profile to toxins that could occur in contaminated shellfish tissue. An antibody-based kit that can detect STX in shellfish is commercially available (RIDASCREEN®), and a second rapid dipstick kit (MIST Alert™) is available for the qualitative detection of positive and negative samples. The latter test is accepted by the United States Food and Drug Administration (US FDA) for use in the United States National Shellfish Sanitation Program. This kit was found effective in all the samples tested in Europe and was applied recently in a survey of Scottish shellfish.

5.4.2. Amnesic Shellfish Poisoning (ASP) Toxins

ASP is caused by consuming bivalve molluscs that have accumulated toxin by feeding on toxic diatom, Nitschia. This is the only shellfish poisoning caused by diatom. Toxin responsible is domoic acid.

Symptoms

Symptoms vary from slight nausea and vomiting to loss of balance, neurological disturbance resulting in confusion and short term memory loss. Tolerance level (as domoic acid) is 20mg/kg.

The parent compound of this group is domoic acid (DA), a polar glutamate agonist. Since the original episode of human poisoning following consumption of DA-contaminated shellfish in Canada in 1987, both the toxin and the diatoms capable of producing it have been found to be distributed worldwide.

Origins and Distribution

The cosmopolitan diatom Pseudonitzschia multiseries was identified as the culprit organism in the 1987 poisoning event. Surveys by a number of groups have identified over a dozen species of Pseudonitzschia capable of producing DA, but only three strains (*P. multiseries, Pseudonitzschia australis,* and *Pseudonitzschias eriata*) are considered to be high producers of DA (>10 pg/cell). Levels of DA inthe other strains are generally <1 pg/cell. The original 1987 event was followed some years later by a more widespread episode along the western coast of the US and today toxin-producing strains of Pseudonitzschia have been found in the Gulf coast of the US and along the western shores of the US and Canada. To a great extent, the wide distribution of toxin-producing strains of the diatom have resulted in the frequent occurrence of DA in many filter feeding organisms such as mussels, scallops, and clams. Another factor is the movement of the toxin through the food chain, which has been well documented in studies along the west coast of the US.

Toxicology

DA is an acidic amino acid and potent neurotoxin unaffected by cooking or steaming. Indeed, it has been reported that cooking or steaming contaminated mussels merely has the effect of spreading the toxin to other tissues of the shellfish as a result of disruption of the digestive gland. When DA was identified as a new marine toxin it was quickly determined to be a glutamate agonist that binds to a sub-class of neuronal receptors called kainite receptors, and review of the toxicologic pathology of DA has been published. The oral absorption of DA is around 5 - 10 per cent of the administered dose, and is mainly distributed in the blood. Penetration of the blood brain barrier is poor in healthy subjects, but any impairment of the blood brain barrier results in additional risk, as does impaired renal function which increases serum concentration and residence time of the toxin. Victims display a variety of symptoms including headache, nausea, seizures, disorientation and coma.

Detection Methods

DA is water soluble and can be extracted from shellfish tissue by the acidic PSP extraction method although this may result in some decomposition. Alternatively, shellfish tissue can be efficiently extracted with aqueous methanol (1:1), and this is typically the method of choice. Provided the concentrations in shellfish are high enough (>50 μg/g shellfish tissue), DA can be detected by the mouse bioassay but the regulatory limit for shellfish is set below this level at 20 μg/g (20 mg/kg) shellfish tissue. Fortunately DA is conveniently detected by a variety of LC-linked analytical methods and this is the recommended approach for DA monitoring in EU countries. The sensitivity of the LC/UV (ultra violet) method (detection limit ~ 1 μg/g) can be improved (20 - 30 ng/g) with a rapid clean-up step using a strong anion exchange resin, as can LC/MS.

5.4.3. Diarrhetic Shellfish Poisoning (DSP)

DSP is caused by eating bivalve molluscs which have accumulated toxin by feeding on bloom of dinoflagellates such as Dinophysis and Procentrum. Toxin responsible for DSP is okadaic acid and its derivatives.

Symptoms

Acute diarrhea with vomiting and abdominal pains. Victims recover within 3-4 days. No mortalities have been reported. Tolerance level is 20 μg/100g. The active agents responsible for DSP incidents are a group of polyethers that include okadaic acid (OA), dinophysistoxin-1 (DTX-1), and dinophysistoxin-2 (DTX-2). In the context of this chapter, the term "DSP toxins" will only be used to refer to these three polyether compounds and any derivative based on them. Since their first characterisation it was believed that the only causative agents of this group were OA and DTX-1, but just over 15 years ago a third toxin, DTX-2 was discovered in contaminated Irish shellfish. Since then, DTX-2 has been found to be extremely widespread throughout many European maritime countries and is often the major DSP toxin present. DSP toxins are produced by dinoflagellates of the genera Prorocentrum and Dinophysis. The benthic Prorocentrum spp. are amenable to laboratory culture and consequently are used in many experiments involving studies of DSP toxins. However, the heterotrophic dinoflagellates Dinophysis spp. are the more potent source of toxins – often very low cells/ml of seawater can result in human poisoning, and consequently were often overlooked as the real source of the toxins. To further complicate matters, Dinophysis spp., produce a variety of other lipophilic compounds that are toxic in the mouse bioassay, but do not have the same mechanism of action as DSP toxins nor induce the same symptoms. Lately these compounds, which include the pectenotoxins and the yessotoxins, have been referred to as lipophilic shellfish poisoning toxins (LSPs or LSTs) because they are found in the same lipophilic fractions as the DSP toxins and are described in more detail later in this chapter. The parent acids are the only DSP congeners to be found in shellfish tissue, although occasionally fatty acyl esters linked through the 7-OH group (referred to as DTX-3 compounds) are also found and are likely a result of bioconversion in shellfish tissue. In comparison, analysis of Prorocentrum cultures have uncovered an additional series of diol esters of OA and sulphated diesters (called DTX-4, DTX-5, and DTX-6) all of which are rapidly converted to OA when the plankton cells are ruptured. While these esters are not toxic in vitro, they can be converted to the parent toxins by non-specific esterases or lipases either in the producing organism or in the host shellfish.

Origins and Distribution

The first member of the group was identified in contaminated Japanese shellfish in 1985 and DSP toxins continue to be a threat in those areas. In this case Dinophysis was suspected as the culprit organism. This was probably not the first DSP episode, and anecdotal evidence now suggests that several seafood poisoning episodes during the 1960s and 1970s in the Netherlands, Norway and

Japan were likely due to DSP toxins produced by a Dinophysis spp. Today over ten Dinophysis spp. are recognised as DSP toxin producers, and include Dinophysis acuminata, Dinophysis acuta, Dinophysis fortii, Dinophysis caudata, and Dinophysis norvegica among others. DSP toxins produced by benthic Prorocentrum spp., such as Prorocentrum lima, Prorocentrum maculosum, and Prorocentrum belizeanum, are also recognised as a threat to human health. Perhaps as a consequence of these multiple sources of DSP toxins, DSP episodes have been reported worldwide and are included in many monitoring programs

Toxicology

The DSP toxins are potent inhibitors of the serine/threonine phosphatases PP1 and PP2A, both critical enzymes involved in many key metabolic pathways in mammalian cells and consequently affect a host of other biological processes in a cell including tumor promotion. All the DSP toxins have around a thousand-fold greater affinity for PP2A than PP1, although a comparative study of in vitro phosphatase inhibition with mouse bioassay data has shown that OA is about twice as potent as DTX-2. Human symptoms include diarrhoea, nausea, vomiting, and abdominal pain. The onset of symptoms can occur within 30 minutes of consuming contaminated shellfish and generally last about 2-3 days.

Detection Methods

Once again the mouse or rat bioassay is relied upon to detect DSP toxins in contaminated seafood, and while this provides critical seafood safety information, it does not provide information on the DSP toxin profile, or indeed on the presence of other associated LSTs that can occur with DSP toxins and which result in a positive mouse bioassay result (vide infra). At the first Meeting of the EU National Reference Laboratories on Marine Biotoxins and Analytical Methods and Toxicity Criteria in 1996, it was agreed that the established mouse bioassay with an observation time of 24 hours is currently the preferred method for the detection of the acute toxicity of acetone soluble DSP toxins. A regulatory level of 80 - 160 μg OA equivalents/kg of whole shellfish tissue has been implemented in some countries. This regulatory level includes the possibility of LSPs in the DSP fraction being tested. Due to the lipophilic nature of the toxins, some careful clean-up steps are required to avoid such matrix interferences and to simplify data interpretation. Preparation of a fluorescent 9-anthryldiazomethane (ADAM) derivative followed by HPLC offers reasonable sensitivity, but the method is complex and matrix effects as well as interferences with other lipophilic compounds such as the pectenotoxins compromise the method, and a detection limitof 100 μg/kg has been reported. In recent years, most progress in DSP toxin detection has been using new or improved LC/MS methods and a detection limit of 10 μg/kg has been reported, even with the minimal of clean up steps. Immunoassays have been developed for OA detection but an inherent problem of these methods has been the low cross reactivity with DTX-1, and DTX-2, which can lead to an underestimation of the total DSP content in a sample. The 7-O acyl derivatives (DTX-3) of OA that can form in shellfish also show poor cross reactivity and this further exacerbates the problem. DSP-Check is a commercially available ELISA test kit that can detect both OA and DTX-1 with comparable sensitivity, with a claimed detection limit of 20 µg/kg, and another rapid antibody-based test kit for DSP toxins has also been reported. In yet another approach, fluorometric protein phosphatase inhibition assays have been found to perform better than colorimetric assays, with good agreement with the mouse bioassay and LC–based methods. If a European collaborative study of the fluorometric protein phosphatase inhibition method is successful it may eventually be approved for regulatory purposes in the EU.

5.4.4. Neurotoxin Shellfish Poisoning (NSP) Toxins

NSP is caused by consuming shellfish that have been exposed to toxic dinoflagellate bloom of Ptychodiscus breve (Gmnodinium breve). Toxin responsible for NSP is a family of brevitoxins which are llipophilic, relatively insoluble in water and soluble in non-aqueous solvents.

Symptoms

Resemble those of PSP except paralysis. It is not fatal and causes neurological symptoms. **Tolerance level:** 20 mu/100g (MU: mouse units). NSP episodes are caused by a family of fused polyether compounds known as the brevetoxins. There are two skeletal types known as PbTx-1 (or brevetoxin A) and PbTx-2 (brevetoxin B). A complicating factor both in terms of establishing the profile and toxicity of brevetoxin-contaminated shellfish has been the discovery of the so-called "brevetoxin metabolites". These are biotransformation products of principally PbTx-2 by addition of certain amino acids to the conjugated aldehyde function.

Origins and Distribution

The brevetoxins are produced by the dinoflagellate Karenia brevis (formerly known as *Gymondinium breve* and *Pytochodiscus brevis*) a dinoflagellate found in the Gulf of Mexico, and off the coast of New Zealand. Interestingly, the same toxins have been reported to be produced by three genera of Raphidophytes, namely Chatonella marina, Fibrocapsa japonica, and Heretosigma akashiwo, which are found in Japanese waters. Many of these species are associated with massive fish kills, and recently the major toxin associated with an ichthyotoxic unialgal bloom of Chatonella cf. verruculosa off the mid Atlantic coast of the U.S. was chemically characterised as PbTx-2. In the Gulf of Mexico, brevetoxins are found in mussels, oysters, whelks, and cockles although recently it has been shown that herbivorous fish can store brevetoxins in their flesh.

Toxicology

Brevetoxins bind to site 5 of voltage-gated sodium channels, which can result in a cascade of associated events in a cell. The human health implications following exposure to NSPs have been reviewed. Ingestion of brevetoxins results in a variety of symptoms including parathesia, vertigo, malaise, nausea, and diarrhoea, and in severe cases seizure. The mean time to onset of symptoms is around 3 hours, with a mean duration of symptoms of around 1 day. Exposure to brevetoxins can also occur through inhalation of brevetoxin aerosols resulting in irritation of the upper respiratory tract. The brevetoxin metabolites are reported to be less toxic than the parent compound.

Detection Methods

The mouse bioassay is used as the primary detection method, and an action level has been set at 800 µg/kg (or 80 µg PbTx-2 equivalents/100 g or 20 MU/100 g or 4 µg/mouse) of shellfish tissue to trigger regulatory action. While the mouse bioassay continues to be used, considerable progress has been made in the development of LC-MS methods to routinely monitor for brevetoxins in shellfish. In this application, pre-treatment or clean-up of the shellfish sample is essential in order to obtain reproducible results, but this method is now used routinely in New Zealand and the US to monitor for the presence of brevetoxins. A second generation competitive ELISA assay has been developed that has successfully quantified brevetoxins in a variety of biological and environmental matrices including seawater, shellfish homogenates, and mammalian body fluids without pretreatment or dilution. In a comprehensive multi laboratory evaluation of five different detection methods as possible replacements for the mouse bioassay, the competitive ELISA method correlated most closely with the mouse bioassay results, and it has also been used in studies to quantify occupational and recreational exposure to aerosolised brevetoxin exposure.

5.5. Ciguatera Fish Poisoning (CFP)

1. CFP is caused from the ingestion of variety of tropical and subtropical carnivorous reef fishes such as Barracuda, Groupers, Seabass, Snappers *etc.*
2. There are over 300 fish species responsible for CFP.
3. The herbivorous reef fishes feeding on toxic dinoflagellates (Gambierdiscus toxicus) are in turn fed by carnivorous fishes which accumulate toxins in their tissue.
4. Toxin accumulation is more in liver followed by viscera and muscle tissue.

Symptoms

- ☆ Upon ingestion of toxin containing fishes, symptoms occur within 3-6 hours.
- ☆ Symptoms are similar to PSP with gastrointestinal disturbances and neurological disorders. These include vomiting, diarrhea, tingling and burning sensation in mouth, lips and throat, muscle cramping and weakness.
- ☆ Death generally occurs due to respiratory failure.

Levels as low as 1ppb in fish can cause illness.

Known for centuries and originally referred to as ciguatera by Spanish explorers in the Indo-Pacific ocean and Caribbean, this family of toxins has proven to be a challenge both to identify and monitor. Outbreaks of ciguatera poisoning are confined to tropical and sub-tropical regions of the ocean, with hot spots in the coral reef areas of the Caribbean and French Polynesia. Ciguatera toxins (CTXs) are extremely potent and occur in finfish rather than shellfish, beginning with herbivores that feed on benthic dinoflagellates associated with reefs and which aretransferred through the food chain to the larger carnivores. CTXs are large, complex, lipid soluble polyethers, and only a few structures have been identified, but more are suspected to exist, most likely a consequence of biotransformation of the parent toxins as they pass up the food chain. Indeed even the chemistry of CTXs from the Pacific differs subtly from those found in the Caribbean. The diversity of benthic phytoplankton that exist on reefs further complicates the picture. Culprit species producing CTXs may simultaneously produce unrelated toxins such as maitotoxin, and other genera of neighbouring reef dinoflagellates can further contribute to an extremely complex profile of toxins in finfish.

Origins and Distribution

It is recognised that several species of benthic dinoflagellates can produce CTXs and the gambiertoxins, a group of closely related polyethers that are considered the source of ciguatoxins. These include species belonging to the genera Ostreopsis and Coolia, although the dinoflagellate Gambierdiscus toxicus isconsidered to be the major source of these toxins. Such microalgae contribute to the primary production of tropical and sub-tropical reefs and live in epiphytic association with bushy red, green and brown seaweeds and also occur in sediments and coral debris. Distribution through the food chain begins with grazing by herbivorous fish including the sturgeon fish and parrotfish, which in turn are consumed by carnivores such as snappers, groupers, barracudas, sharks, and moray eels. Local island populations attempt to avoid poisoning by catching only smaller fish, but fish toxicity can vary widely depending on where the fish are caught and how long they have been feeding in that area. Nevertheless CTX poisoning continues to be a threat to local and visiting populations, and there are suggestions that such incidents are increasing and spreading worldwide.

Toxicology

Ciguatoxins (and maitotoxin which is often co-produced with CTXs) are extremely poisonous, although not all CTXs are equipotent, and it is estimated that ingestion of as little as 0.1 μg of ciguatoxin can result in human illness. Like the brevetoxins, they bind to voltage-dependant sodium channels in the cell membrane causing them to open, and inducing membrane depolarisation resulting in prolonged symptoms indicative of damage to nerve tissue. Differences in potency among the CTXs correlate directly with their binding affinity for the sodium channel. The pharmacological effects following CTX exposure are numerous and include severe gastrointestinal distress and neurological symptoms, all of which can be linked to the effect of CTX on excitable membranes. The onset of symptoms can occur rapidly within 24 hours, and although the gastronomical problems may dissipate within a day or so, the neurological symptoms can persist for days, weeks or even months. These include paresthesias (numbness and tingling of the extremities), and temperature sensation reversal, where cold objects feel hot and vice versa. Other symptoms include a metallic taste, pruritus, arthralgia, myalgia, and the sensation of loose teeth. Another symptom that can develop over time is extreme fatigue, which can be confused with chronic fatigue syndrome. Patients have reported symptoms lasting weeks and even months, which can return following the consumption of alcohol and certain foods such as oily fish.

Detection Methods

As a consequence of their potency, only trace amounts of CTXs can cause illness following consumption of contaminated fish, and in fact CTX poisoning is often only identified after consumption of contaminated fish. Even when suspect samples are available, the absence of a strong chromophore means that simple detection methods based on UV or fluorescence-based detection methods are ineffective. Globally, there are neither standards nor an official testing program, though the mouse bioassay continues to be relied upon as the most common method of detecting CTXs. The recommended method to extract fish tissue has been described in detail and the initial warm acetone extract is partitioned against hexane and diethyl ether where the CTXs are concentrated. A good knowledge and careful observation of the symptoms in the bioassay (Minimal Lethal Dose (MLD) 0.25 μg/kg) can provide useful clues as to the toxins present but more specific and informative assays are required. Immuno assays may be an alternative detection method, although specificity and sensitivity remain major hurdles and the presence of DSP toxins affects the test. A sodium channel-specific assay using mouse neuroblastoma cells has been described although this approach has still to be developed in a routine, high-throughput screen. Recently, another *in vitro* sodium channel-based assay, specific for CTX activity, has been developed, but this only has application in a clinical setting. Chemical detection methods are severely compromised by the lack of standards, and the separation of these polyethers from the lipid-soluble fraction of fish provides an additional challenge. Consequently, only a few laboratories are able to undertake such analyses, which are based on an LC-MS/MS approach. Acetone extracts of fish tissue are subjected to solvent partitioning, followed by clean-up on silica and amino propyl cartridges. This is then subjected to LC-MS/MS analysis, which features characteristic fragment ions corresponding to sequential losses of three molecules of water. In this way Caribbean and Pacific CTX-1 compounds can be distinguished.

5.6. Tetrodotoxin/Pufferfish Poisoning

1. Puffer fish poisoning is caused due to the consumption of puffer fish (tetradon fish or fugu) which have toxic tissues or organs.

2. Pufferfish toxin is thought to be produced by the symbiont bacteria (Pseudomonas) associated with the fish.
3. Tetradotoxin is chemically amino per hydro quinazoline which is similar to saxitoxin of PSP and cause symptoms similar to PSP but of varying degree.
4. Only certain species of puffer fish are toxic and toxin is restricted to the skin, liver, viscera, gonads, intestine and muscle.
5.. Since the fugue is a delicacy in Japan, most cases of intoxication are reported from Japan with instances of death.

Symptoms

Causes neurological symptoms similar to PSP – tingling in lips and extremities, paralysis and death by respiratory arrest and/or cardiovascular collapse. Mortality rate is high. Cardiovascular effects are more severe than PSP with high death rates.

Tetrodotoxin (TTX) possesses a unique chemical structure and was the first marine toxin to be chemically characterised. Since its discovery, other derivatives have been reported, and over ten analogs are known.

Origins and Distribution

Unlike many other marine toxins there is no temporal distribution of TTX episodes. Most intoxication events have occurred in South East Asia, including Japan, Thailand, Taiwan, China, as well as other parts, and islands of the South Pacific. TTX most commonly occurs in many species of puffer fish, specifically in the liver, gonads and roe. In Japan, where puffer fish of the genus Fugu are a delicacy, it is served in speciality restaurants with qualified chefs to properly prepare the fish. However TTX has been found to occur in a remarkable variety of genetically unrelated species including gastropods, the blue-ringed octopus, molluscs, horseshoe crabs, and even newts, raising speculation that TTX is produced by a bacterial source. This was indeed confirmed and a number of marine Vibrios have been shown to be capable of TTX production. Since then, this has expanded to include bacteria within the Alteromonas and Pseudomonas genera, and more recently by an actinomycete Norcardiopsis dassonvillei.

Toxicology

Like the PSP toxins, TTX acts on the central nervous system by binding extracellularly to site 1 of voltage-gated sodium channels, and blocks diffusion of sodium through the sodium channel, preventing depolarisation and propagation of action potentials in nerve cells. All of the observed toxicity is secondary to the action potential blockade. The onset of symptoms usually occurs rapidly, and in mild cases results in distal muscle weakness, nausea and vomiting. Severe poisoning causes paralysis and respiratory failure, and ingestion of larger amounts leads to cardiovascular effects including bradycardia, hypotension, respiratory failure and coma. An oral dose of 1 - 2 mg of purified toxin can be lethal.

Detection Methods

TTX can be extracted from suspect tissue using dilute acid conditions, and once again the mouse bioassay has been used to detect the presence of the toxin. A review of TTX detection methods has been published. Chemical methods for TTX detection have also been investigated and are mainly based on its smooth conversion by heating in an alkaline solution to a detectable fluorescent (C_9) base. Typically this has been accomplished using HPLC followed by postcolumn oxidation to the fluorescent derivatives. However this method has been criticised for being complex, and more rapid LC-MS methods [(M+H)+ m/z 320] have been reported for the detection and identification of TTX and its various derivatives. Immunoassay methods have also been investigated, and while promising, need further development.

5.7. Azaspiracids

Azaspiracid (AZA1) was originally identified in 1995 as the causative agent for a poisoning episode in Ireland, following consumption of contaminated mussels Mytelius edulis. Recent reviews of this new group of toxins have appeared, and another dealing with the chemistry and detection methods has been published. Since the original discovery and identification of the toxin, almost 20 additional derivatives have been reported. The majority of these have been proposed based on LC/MS analytical data, and only a few structures (AZA 1 - 5) have been fully characterised by Nuclear Magnetic Resonance (NMR) and MS analysis. Usually the major AZA derivatives found in shellfish are AZA1 (50 - 80 per cent), with AZA2 (10 - 30 per cent) and AZA3 (5 - 20 per cent) present in lesser amounts. Usually the additional AZA derivatives found in shellfish tissue are hydroxylated derivatives (*e.g.* at C-3 and C-23), and it is likely that most of these compounds are biotransformation products formed in the shellfish tissue.

Origins and Distribution

The azaspiracids were first isolated from contaminated shellfish, and to date this has been the only source from which these toxins have been found at toxic levels. However, as in other cases of shellfish toxins, these molluscs were never considered to be the biogenetic source. There has been considerable debate over the source of azaspiracids, and some time after the original episode in Ireland, it was reported that the heterotrophic dinoflagellate Protoperidinium cassipes was the planktonic source of the toxins. While this result is not necessarily in debate, it has been impossible to establish that P. cassipes produces AZAs in culture, and variable field results have led to suggestions that the actual producing organisms are Dinophysis spp. which are ingested by P. cassipes. In fact the routes of trophic transfer of AZAs through the food chain ending in contaminated shellfish remain unknown. Following the original episode, AZA occurrence has been reported by several European countries including France, Portugal, Scandinavia, and off the coast of Morocco. The bivalves affected include mussels (*Mytilus edulis,*

Mytilus galloprovincialis), oysters (*Crasso streagigas, Ostrea edulis*), scallops (*Pecten maximus*), clams (*Tapes philipinarum, Ensis siliqua, Donax* spp.), cockles (*Cerastoderma edule*), and most recently in a crustacean (*Cancer pagurus*).

Toxicology

Consumption of AZA contaminated shellfish can result in acute symptoms that include nausea, vomiting, diarrhoea and stomach cramps, which can persist for 2 - 3 days, and to date no long-term effects have been recorded. These symptoms are not unlike those experienced following the consumption of DSP toxins, but current data indicates that unlike the DSP compounds, AZA toxins donot inhibit the phosphatases PP1 and PP2, though the possibility exists that they may inhibit other phosphatases. The limited amounts of AZA toxins available continue to hinder a complete understanding of their mechanism of action. Injection of a lethal dose of AZA1 in mice (>150 µg/kg) caused vacuole formation and fatty acid accumulation in hepatocytes, parenchymal cell pyknosis in the pancreas, abnormalities in the thymus and spleen, and erosion and bleeding in the stomach. Importantly these pathological outcomes were reported as being different from those caused by DSP, PSP, and ASP. Oral administration of AZAs at substantially higher doses than those used in IP injection studies did not result in any mouse mortalities after 24 hours, although autopsies conducted after 4 and 8 hours revealed various gastrointestinal (GI) abnormalities including accumulation of fluid in the ileum, and necrosis of epithelial cells on the microvilli. Based on IP minimum lethal dose data, it is suggested that the order of toxicity is AZA2 (110 µg/kg) >AZA3 (140 µg/kg) >AZA (150 µg/kg). Using the same approach, it is believed that the hydroxylated derivatives AZA4 (470 mg/kg) and AZA5 (<1000 mg/kg) are considerably less toxic. Once again, the lack of sufficient material and other logistical problems have prevented complete studies of any chronic effects arising from long term exposure to low levels of the AZA toxins. AZAs also display cytotoxicity towards various mammalian cell lines.

Detection Methods

The regulatory level for AZAs in shellfish has been set by the EU at 160 µg AZA equivalents/kg shellfish tissue. The mouse bioassay has been used to monitor shellfish extracts, but unlike DSP monitoring methods which use only the hepatopancreas of the shellfish for extraction, it is recommended that the entire shellfish be used for AZA analysis since it has been shown that these toxins can migrate to other tissues. Currently LC/MS and LC-MS/MS analysis is the preferred chemical analytical detection method for AZAs, which are particularly amenable to detection by positive ion electrospray mass spectrometry. Key fragment ions are useful in identifying the various AZA compounds that can occur in shellfish tissue. However, interferences can arise when analysing a matrix as complex as a crude shellfish extract and it is recommended that a simple solid phase extraction (SPE) step be introduced prior to LC/MS analysis.

5.8. Cyclic Imines

These toxins are represented by several distinct structural types, but all contain a spirocyclic or cyclic imine function within their structure. This common cyclic imine feature is believed to be a common pharmacophore of this group of toxins which comprises the gymnodimines, the spirolides, the pinnatoxins, and closely related ptierotoxins. Due to their relatively recent discovery and limited knowledge of their toxicological properties and impact on human health, they are sometimes referred to as emerging toxins of uncertain human health concern, but they are included here for completeness. Cyclic imines are often referred to as fast acting toxins due to their "all or nothing" effect in the mouse bioassay over a very narrow dose range. In only two cases (gymnodimines and spirolides) has a planktonic source been identified. Reviews of the current status of the cyclic imine group were published by Molgo and Cembella.

Gymnodimines

The observation of neurotoxic symptoms in the mouse bioassay of lipophilic extracts of dredge oysters (*Tiostrea chilensis*) collected from around the South Island of New Zealand, led to the identification of a new marine toxin with an unprecedented structure named gymnodimine. The structure is novel in that it contains a spirocylic imine group and a single tetrahydrofuran ring embedded within a macrocyclic structure. Some time later, two additional congeners; gymnodimine B and C were isolated from the dinoflagellate Kareniaselliformis (formerly *Gymnodinium selliformis*). The gymnodimines are the smallest members of the cyclic imine group with a molecular weight range of 500 - 520 Da.

Origins and Distribution

The original observation of toxicity in oysters led to a wider survey of other shellfish species in New Zealand when it was discovered that the toxin was present in a variety of bivalve molluscs including mussels, scallops, cockles, Pacific oyster, surf clam and New Zealand abalone. At the time of the initial observation of gymnodimine, it was proposed that a dinoflagellate identified as *Gymnodinium* spp., a species closely related to *Gymnodinium mikimoitoi* was the planktonic source of the toxin. Subsequent taxonomic studies revised this again to a new dinoflagellate *K. selliformis*.

Toxicology

A review of the toxicology of all cyclic imines has been published. The original observation of toxicity in New Zealand followed from IP injection of lipophilic extracts in the mouse bioassay. A later study using purified toxin established an LD_{50} dose (Lethal Dose, 50 per cent) of 96 μg/kg. There is a rapid onset of symptoms including respiratory distress, paralysis of the hind quarters and eventually complete immobility, abdominal breathing, and lack of responsiveness to stimuli. The time lapse between ingestion and death is between 5 and 15 minutes. Reduction of the imine group eradicated the toxicity, and short-acting cholinesterase inhibitors such as neostigmine or physostigmine protected mice against the toxic effects of gymnodimine. This led to the suggestion that gymnodimine acts by blocking nicotinic acetylcholine receptors at the neuromuscular junction. Oral administration of the toxin to mice by gavage resulted in about an 8- to10-fold reduction of toxicity (LD_{50} 600 - 900 μg/kg), and there was a further ten-fold reduction in toxicity when the toxin was supplied with food. Based on this diminished oral toxicity, together with a lack of evidence of human intoxication, it has been stated that regulation against gymnodimine is not required.

Detection Methods

The characteristic "fast-acting, all or nothing" mouse bioassay response to gymnodimines and cyclic imines in general continues to be a useful monitoring approach. Most cyclic imines lack an extended chromophore, and so spectroscopic detection by UV methods is not favoured. However, cyclic imines are readily observed by electrospray mass spectrometry in the positive ion mode, and this is the detection method of choice.

Whether isolated from dinoflagellate cells directly or from shellfish tissue, the first step in the process is typically an initial clean-up step using a short C18 cartridge and elution with water containing increasing amounts of methanol, finishing with 100 per cent methanol. Fractions are then dried and re-suspended in methanol or acetonitrile in preparation for LC/MS analysis.

Spirolides

The spirolides comprise the largest group of cyclic imine compounds. They were originally discovered in extracts of mussels (*M. edulis*) and scallops (*Placopectinma gellanicus*) from eastern Canada that displayed abnormal symptoms in the mouse bioassay. In particular, extracts exhibited the fast acting symptoms that became associated with cyclic imines following IP injection. Since their original discovery which resulted in the characterisation of spirolides A-F (99,100), almost a dozen derivatives in total have been identified to date. The structure of the spirolides bears a strong resemblance to both gymnodimine and the pinnatoxins (vide infra). Structural differences among the various spirolides result from variations in the number of methyl groups and the sequence of spiroketal rings 6-5-5 in one group and 6-6-5 in the other.

Origins and Distribution

At the time of the original discovery of the spirolides in shellfish from the coastal waters of eastern Nova Scotia, their localisation in the digestive gland of mussels and scallops led to the suspicion that plankton were the real source of the toxins. This proved to be the case and eventually the dinoflagellate Alexandrium ostenfeldii was identified as the culprit organism. Since then the spirolides have been found to be globally distributed and are found in shellfish or plankton fractions along eastern Canada, the gulf of Maine, in fjords in Denmark and Norway, southern

Ireland, the western coast of Scotland, the Bay of Biscay and most recently from Mediterranean shellfish and plankton in the Adriatic.

Toxicology

The spirolides are the most studied group of the cyclic imines. The LD_{50} values vary considerably depending upon whether the compound is administered by IP injection (5-8 µg/kg), by gavage (87-166 µg/kg; fasted mice), or by oral ingestion (381-707 µg/kg; fasted mice). All data reported here are for desmethyl spirolide C. While oral toxicity is considerably less than by IP injection, those spirolides possessing the vicinal dimethyl grouping of the cyclic imine ring are the most toxic of the group. Interestingly, these are the toxins that are also most resistant to acid hydrolysis of the imine function. While the mechanism of action remains to be clearly defined, experiments have shown upregulation of muscarinic and nicotinic receptors in mice following dosing with spirolide C, consistent with modulation of acetyl cholinergic receptors.

Detection Methods

As discussed for gymnodimine, the mouse bioassay can be used to detect these fast acting toxins. However, more useful information on the amount and distribution of toxins present in a shellfish extract is obtained by LC-MS analysis usually following a simple clean-up step. Once again this relies on the sensitivity of these cyclic imines to detection by positive ion electrospray mass spectrometry. A number of methods following this procedure have been described.

Pinnatoxins and Pteriatoxins

Pinnatoxins A - B contain a spiro imine moiety similar to that found in the spirolides, and in fact these compounds share a number of other structural features. The compounds are zwitter ionic and therefore some what more polar than the spirolides. To date, four members of the group have been identified – two of them being the C-34 diastereomers. The pteriatoxins were isolated from a different mollusc but bear a strong structural resemblance to the pinnatoxins. Three members of this group have been reported, which includes two diastereomers.

Origins and Distribution

Shellfish of the genus Pinna, which occupy tropical waters around Okinawa and the south east coast of China, have long been suspected of causing human illness after ingestion. This prompted a chemical investigation of Pinna muricata that ultimately led to the identification of the pinnatoxins in 1995, and later the closely related pteriatoxins were isolated from the digestive glands of another Okinawan bivalve Pteria penguin. There is strong circumstantial evidence to suspect that these toxins, like the gymnodimines and spirolides, arise from an algal source - most likely a dinoflagellate, but this remains to be conclusively established. However, there is no compelling evidence to suggest that they have resulted in human toxicity.

Toxicology

There is little information describing the toxicology of these compounds. Only the acute toxicity in mice following IP injection has been reported. The LD_{99} (Lethal Dose, 99 per cent) values for pinnatoxins A-C are in the range (22 - 180 μg/kg), while pinnatoxin D was essentially inactive. The LD_{99} dose for the pteriatoxins is reported to be in the range (8 - 100 μg/kg).

Detection Methods

No formal detection methods for these compounds have been reported, but the mouse bioassay data and LC-MS can be used to monitor suspect seafood for these compounds.

Conclusion

There has been considerable progress in the detection of known marine toxins, which in many cases are now found to be more widespread than previously believed. This is generally a result of increased monitoring and continually improving detection technologies and methods. Coupled with this, new toxin groups such as the azaspiracids and cyclic imines have also been reported, once again reflecting improvements in purification and spectroscopic methods permitting structural identification of new and complex structures. Despite this progress, research and certified standards of the key toxins are often unavailable or in short supply and in the future considerable logistical effort will be required to address this issue.

With increasing knowledge of the chemistry of toxins, and more advanced means of detection, has come the realisation that a "toxin profile" can be a complex mixture of many different classes of toxins, and in some cases it is known that one toxin can potentiate or augment the toxicity of the other. While this is recognised as a real possibility for a wide range of toxins, to date action or regulatory levels for toxins continue to be assessed only for individual purified toxins. In most cases this is due to the paucity of material that is usually available for any single group of toxins, never mind the amounts required of several groups, in order to fully study the synergistic effects of multiple toxin types.

While the mouse bioassay continues to be the mainstay of most monitoring programs, an encouraging development is the use of antibody-based methods to detect the presence of marine toxins. The challenge facing such approaches is the variety of matrices that must be examined, and in particular the ruggedness of such methods, but these are gradually being overcome. Even if they do not completely replace other methods such as the mouse bioassay or chemical analysis that require sophisticated chemical analytical instrumentation, they can provide a valuable "yes/no" filter for shellfish and finfish extracts, thus significantly reducing the number of mouse or chemical analyses that have to be performed.

5.9. Scombroid Poisoning

Scombroid poisoning or histamine poisoning is caused by the consumption of fishes containing high levels of histamine. Scombroid fishes (tuna, seer fishes, mackerel) containing red meat are implicated in scromboid pisoning. Other fishes like carangids, herrings, sardines and anchovies are also involved in histamine poisoning.

Source of Histamine

Scombroid fishes have high levels of histidine which is converted to histamine by the growth of microorganisms pocessing the enzyme histidine decarboxylase. Conversion of histidine to histamine by microorganisms results in the accumulation of histamine in fish. Bacteria involved in decorboxylation are *Morganella morgani, Klebsiella pnemoniae, Enterobacter aerogenes* and *Haffnia alvei*. The main source of these bacteria to fish is from post harvest contamination. Though these bacteria grow well at 10°C, highest histamine production occurs at 37°C. Histamine is heat resistant.

Symptoms

1. Symptoms of poisoning occur with short incubation period (few minutes to few hours), the illness is mild and self limiting lasting for only few hours.
2. Gastrointestinal disturbance (nausea, vomiting, diarrhea), facial flushing, labial edema, itching of the skin and rashes on skin are the common symptoms of illness.

Maximum permissible limit in seafood is 50 ppm.

Toxins produced by microorganisms proliferating in aquatic foods are responsible for several foodborne diseases. Histamine toxicity or poisoning is a food borne chemical intoxication caused by consuming fish containing high levels of histamine. The importance of histamine toxicity was recognized by Japanese scientists in 1950's.

What is Histamine?

Histamine is a small, polar, organic molecule of low molecular weight, which comprises an imidazole ring and an ethylamine side chain. It is a produced naturally by the body and is also present in many foods such as beer, cheese, wine, *etc.* In the foods, it is produced from amino acid histidine by the removal of carboxyl group (COOH) by the enzyme, histidine decarboxylase. It is a biologically active amine, which has important physiological effects in humans, either psychoactive or vasoactive.

Histamine Production in Fish

During spoilage, certain bacteria produce decarboxylase enzymes, which act on free histidine and other amino acids in fish muscle to form histamine and other biogenic amines such as putrecine, cadaverine and spermine and spermidine.

Pathways for the Formation of other Biogenic Amines

Amino Acid Precursor	*Amino Acid Decarboxylase*	*Biogenic Amines*
Tyrosine	Tyrosine decarboxylase	Tyramine
Tryptophan	Tryptophan decarboxylase	Tryptamine
Lysine	Lysine decarboxylase	Cadaverine
Phenylalanine	Phenylalanine decarboxylase	Phenylethylamin e
Arginine	Arginine decarboxylase	Agmatine
Ornithine	Ornithine decarboxylase	Putrescine
		Spermidine
		Spermine

Histamine Toxicity in Aquatic Food Products

The foodborne intoxication caused due to the consumption of fish containing high levels of histamine and/or other biogenic amines was originally called as "scombroid poisoning," because it was associated with the consumption of fish belonging to the families Scombridae and Scomberesocidae. However, this term was later found misleading, as fish from non scombroid families are also implicated in this poisoning and hence called as "histamine poisoning" or "histamine toxicity".

Fish that commonly cause histamine poisoning includes scombroids like tuna (*Thunnus* spp.), skipjack tuna (*Katsuwonus pelamis*), mackerel (*Scomber* spp.), saury (*Cololabis saira*) and bonito (*Sarda* spp.) and non scombroids like mahi mahi (*Coryphaena* spp.), sardines (*Sardinella* spp.), pilchards (*Sardina pilchardus*), anchovies (*Engraulis* spp.), herring (*Clupea* spp.), marlin (*Makaira* spp.) and blue fish (*Pomatomus* spp.), Western Australian Salmon (*Arripis truttaceus*), sockeye salmon (*Oncorhynchus nerka*), amberjack (*Seriola* spp.) and cape yellow tail (*Seriola lalandii*).

The aquatic food products that are reported to have high histamine levels are canned products of tuna, sardines, pilchard, anchovies and mackerel, semi preserved anchovies, fermented and cured fish products like cured sardine, anchovy and mackerel; fermented fish sauce, shrimp sauce, fish paste, rice bran pickles of sardines, whole sundried anchovies and sundried scad.

Clinical Aspects of Histamine Toxicity

The most common symptoms of histamine toxicity involve

1. Eyes itch, burn or become watery
2. Nose itches, sneeze and produce more mucus
3. Skin itches, develops rashes or hives
4. Sinuses become congested and causes head aches
5. Lungs wheeze or have spasms
6. Stomach experiences cramps and diarrhoea

7. Oedema (swelling) usually manifested as acute urticaria (rapidly appearing hives, accompanied by severe itching)
8. Burning sensation or peppery taste in the mouth
9. Facial flushing, tachycardia, pruritis and asthma like symptoms

Histamine toxicity produces one or more of the above symptoms and is usually a mild illness with short duration, which makes treatment unnecessary. For severe cases or in patients with underlying medical conditions, oral antihistamines are beneficial. Illness begins minutes to hours after ingestion of toxic fish. The lack of severity may be because histamine is poorly absorbed from the gastro intestinal tract, the liver and intestinal mucosa has more capacity to detoxify histamine. The detoxification system is composed of enzymes such as diamine oxidase and histamine-N- methyl transferase. These enzymes convert histamine into non-toxic products. This detoxification system is adequate for handling normal dietary intakes of histamine. However, it fails to detoxify large amounts of histamine when fish containing high histamine content is ingested.

Potentiaters of Histamine Toxicity

Histamine does not appear to be the only causative agent of histamine toxicity since it is not itself toxic when taken orally. The paradox between lack of toxicity of pure histamine and the apparent toxicity of even smaller doses of histamine in spoiled fish has been attributed to the possible occurrence of histamine toxicity potentiaters in spoiled fish. These potentiaters decreases the threshold level of histamine needed for histamine poisoning symptoms in humans challenged orally. Food borne substances that have been suggested as potentiators of histamine poisoning comprise trimethylamine, trimethylamine oxide, agmatine, putrescine, cadaverine, anserine, spermine and spermidine. Apart from diamines, the influence of marine toxins (like saxitoxin and polyether diarrhetic shell fish poisons) on histamine toxicity is also reported.

Two theories have been put forward to explain the potentiation of histamine toxicity. The first one relates to potentiation via enzyme inhibition, where potentiaters inhibit the primary histamine metabolizing enzymes such as diamine oxidase (DAO) and histamine-N- methyl transferase (HMT). As per the second theory, popularly called as barrier disruption theory, potentiaters may inhibit the protective action of intestinal mucin. The intestinal mucin is known to bind histamine in vitro and this binding is essential to prevent passage of histamine across the intestinal cell wall. Potentiaters such as putrescine and cadaverine may bind competitively and preferentially with mucin facilitating the movement of histamine into circulation.

The drug monoamine oxidase inhibitor (MAO I) potentiates the toxicity of histamine. Some of the medications of anti tuberculosis (such as, isoniazid), anti-malarial, anti-trypanosomal drugs contain the DAO inhibitor component. Apart from the medications, other factors such as alcohol consumption and gastro intestinal diseases also aggravates the toxicity problem.

Therapy

Antihistamine drug is the optimal mode of therapy for histamine poisoning. Symptoms usually subside rapidly after such treatment. Both H1 antagonists (*e.g.*, diphenlhydramine) and H_2 antagonists (*e.g.*, cimetidine) have been used for the treatment of histamine poisoning. Since the adverse responses are self-limited, pharmacological intervention may not be necessary in mild cases.

Microorganisms Involved in Histamine Formation

Histamine formation in food requires available amino acids and amino acid decarboxylases synthesized by bacteria. Histamine is formed in fish by certain microorganisms capable of producing the enzyme histidine decarboxylase. Only a few bacteria have the capacity of prolific histamine formation during spoilage. The histidine decarboxylases produced by these bacteria catalyze the conversion of free histidine or histidine generated during proteolysis in the muscle of fish to histamine.

Gram-positive and Gram-negative bacteria can both produce the enzyme, histidine decarboxylase. In fish, histamine decarboxylase enzymes are generally found in bacteria belonging to the family Enterobacteriaceae. Enteric bacteria such as *Morganella morganii, Raoultella planticola, Hafnia alvei, Proteus vulgaris, Enterobacter aerogenes, E. agglomerans, E. cloaceae, Proteus mirabilis, Citrobacter freudii, Escherichia coli, Providencia stuartii, Klebsiella oxytoca, Edwardsiella, Serratia liquifaciens* and *S. marsescen* possess histidine decarboxylases. Among these bacteria *M. morganii*, has the strongest histidine decarboxylase activity. *K. pneumonia, H. alvei* and some strains of *E. cloaceae* and *E. aerogenes* produce more than 500 mg/kg of histamine depending on storage temperature. Non enteric bacteria such as *Clostridium perfringens* is also reported to be most active during decomposition stages of fish. Gram positive bacteria such as *Acinetobacter, Aeromonas, Staphylococcus* and *Bacillus* spp. also has potential to produce histamine by decarboxylase activity. In the case of fermented seafood, *Staphylococcus* spp. and *Tetragenococcus* spp. are reported as histamine producers.

Sources of Histamine Forming Bacteria

In fish, biogenic amine-producing bacteria are most likely to be present on the gills, skin, or in the gastrointestinal tract. Transfer of these bacteria to the flesh of fish, where free amino acids are present, leads to the development of histamine and other biogenic amines. Transfer can occur from the gastrointestinal tract after harvest, through migration, or rupture or spillage of gastric contents during gutting. The amount of amines produced depends on the level of free amino-acids present, which is related to species of fish, and the amount and activity of decarboxylase enzymes. The quantity of decarboxylases is related to the number of decarboxylase producing bacteria transferred to the fish and on the extent to which they multiply.

Factors Affecting the Growth of Histamine Forming Bacteria

Many conditions can affect growth of histamine forming bacteria. Temperature is the main determinant. The concentrations depend on the combined influence of time and temperature: which includes temperature abuse in fishing vessels, inadequate chill storage procedures, inadequate freezing and thawing procedures, temperature abuse in the preparation of dried/smoked products, temperature abuse of tuna for sashmi market, poor canning procedures, *etc.* Other important factors are salt, oxygen availability, type of fish, competition with other spoilage microorganisms and post catching contamination.

1. Temperature

Histamine forming bacteria are capable of growing and producing histamine over a wide temperature range. Growth of histamine is more rapid, however, at high-abuse temperature (21.1°C) than at moderate-abuse temperatures (7.2°C). Growth is particularly rapid at temperature near 32.2°C. Histamine formation more commonly occur as a result of high temperature spoilage than long-term, relatively low-temperature spoilage. The enzyme can be active at or near refrigeration temperatures. The enzyme remains stable while in the frozen state and may be reactivated very rapidly after thawing.

2. Processing Methods

Freezing may inactivate some of the histamine forming bacteria. Cooking on the other hand inactivates both the enzyme and the bacteria. However, once histamine is produced, it cannot be eliminated by heat (including retorting) or freezing. After cooking, recontamination of the fish with the histamine forming bacteria is possible leading to additional histamine formation.

Some of the histamine-forming bacteria are halotolerant or halophilic. Some are more capable of producing histamine at elevated acidity (low pH). As a result, histamine formation is possible during processes such as brining, salting, smoking, drying, fermenting, and pickling until the product is fully shelf-stable. As a number of histamine forming bacteria are facultatively anaerobic, vacuum packaging, modified atmospheric packaging and controlled atmospheric packaging could not control histamine formation. Histamine could also be present in products such as fish protein concentrate that are prepared from muscle or aqueous components of fish tissue.

3. Harvesting Methods

In some harvesting practices, such as long lining and gill netting, death may occur many hours before the fish is removed from the water. Under the worst conditions, histamine formation can already be underway before the fish is brought onboard the vessel. This condition can be further aggravated with certain tuna species that generate heat, resulting in internal temperatures that may exceed environmental temperatures and increasing the likelihood of conditions favourable to growth of enzyme-forming bacteria.

Controlling Histamine Formation

Rapid chilling of fish immediately after death is the most important element in any strategy for preventing the formation of histamine, for fish that are exposed to warm water or air, especially for tunas, which generate heat in their tissues. Some recommendations to control histamine toxicity as per USFDA guidelines are as follows:

1. Fish exposed to air or water temperatures above 28.3°C should be placed in ice, or in refrigerated seawater, ice slurry, or brine of 4.4°C or less, as soon as possible after harvest, but not more than 6 h from the time of death
2. Fish exposed to air and water temperatures of 28.3°C or less should be placed in ice, or in refrigerated seawater, ice slurry, or brine of 4.4°C or less, as soon as possible after harvest, but not more than 9 hours from the time of death; or
3. Fish that are gilled and gutted before chilling should be placed in ice, or in refrigerated seawater, ice slurry, or brine of 4.4°C or less, as soon as possible after harvest, but not more than 12 hours from the time of death; or
4. Fish that are harvested under conditions that expose dead fish to harvest waters of 18.3°C or less for 24 hours or less should be placed in ice, or in refrigerated seawater, ice slurry, or brine of 4.4°C or less, as soon as possible after harvest, but not more than the time limits listed above, with the time period starting when the fish leave the 18.3°C or less environment.

Methods for Histamine Analysis

A number of analytical methods have been developed to detect the level of histamine. They range from bio assays, semi quantitative, quantitative and qualitative techniques to rapid methods and bio sensors that can detect or determine the level of histamine depending upon their sensitivity limit. The methods followed are bio assay-guinea pig ileum contraction, Daphnia assay, Thin layer chromatography, Gas chromatography, HPLC, Ion pair chromatography, Spectroflourometry, Flow injection analysis, calorimetric method, Ion exchange separation, rapid enzymatic method, Enzyme electrode, test strips, dipstick test and capillary zone electrophoresis. However, the flourometric procedure is the official AOAC method commonly used for the determination of histamine.

Regulations

Histamine toxicity has been a food safety concern despite the high hygienic standards imposed by several legislation and guidelines. The limits on the allowable levels of histamine in fish and aqua products vary among nations (Table). The USA has set a lower limit based upon the presence in its populations of individuals with a high sensitivity to histamine.

Limits of Histamine in Fish and Fishery Products

Standard/Country	*Product*	*Limit*
USFDA	Tuna, mahi-mahi and related fish	500 ppm based on toxicity 50 ppm defect action level
European Union (EC No. 2073/2005)	Fishery products from fish species associated with high amount histidine (species belonging to families of Scombridae, Clupeidae, Engraulidae, Coryfenidae, Pomatomidae and Scombresosidae	n=9, c=2, m=100ppm; M=200 ppm
EU No. 1019/2013	Fish sauce	400 ppm max
India (FSSR, 2011)	All fish	200 ppm max
Russia (Sanitary and epidemiological rules and norms, SanPiN 2.3.2. 1078-01)	Tuna, Mackerel, Salmon and Herring	100 ppm max
Australia	Scombridae (tuna and mackerel), Clupidae (herrings, sardines *etc.*), Engraulidae (Anchovy), Coryphaenidae (dolphin fish)	200 ppm, max
Canada	Enzyme ripened products (e.g. anchovies, anchovy paste, fish sauce)	20 mg/100 g max.
	Scombroid fish products (e.g. canned or fresh or frozen tuna, mackerel, mahi mahi)	10 mg/100 g max.

5.10. Faecal Indicator Organisms

Metabolic Products as Indicators of Quality

Metabolic products of microorganisms may be used to assess and predict microbial quality in some products. In fishes, the production of nitrogenous compounds by spoilage organisms has been used as indicators of product quality. These include;

1. Trimethy amine (TMA)
2. Total volatile bases (TVB) such as ammonia and dimethyl amine,
3. Total volatile nitrogen (TVN).

Increase in their value indicates the spoilage, and their concentration increases as the spoilage

5.10.1. Indices of Fish Sanitary Quality

Quality of the fish refers to wholesomeness of fish making it suitable for human consumption. Quality depends and determined by physical, chemical, organoleptic.

Quality Indicators

From the microbiological point of view, the product quality (shelflife) can be assessed by determining the presence of certain microorganisms as indicators of quality, presence of metabolic products of microorganisms and total viable counts. This information will help not only in assessing the existing quality of product but also in predicting product shelflife.

Presence of Certain Microorganisms as Indicators of Quality

The microbiological quality of a product depends on the associated spoilage organisms whose increasing numbers results in loss of product quality, and human pathogenic microorganisms whose presence leads to health risk to consumers. Presence of pathogenic microorganisms can be assessed by detecting an organism called indicator organism which can indicate likely presence of pathogens in food. Thus, the overall microbial quality of the product is a function of number of organisms, and therefore shelflife can be increased by their control.

Microbial indicators are more often used to assess food safety and sanitation rather than quality. Among the several food associated microorganisms the most commonly used organism as indicator is feacal coliform bacteria, *E. coli* and fecal streptococci. These microorganisms are associated with human faeces and their presence in food indicates poor sanitation, likely presence of other pathogens and fecal contamination.

Characters of Indicator Organisms

1. Present and detectable in all foods whose quality to be assessed.
2. Their growth and numbers should have a direct negative relation with product quality.
3. Have history of constant association with pathogen and their numbers should correlate with pathogen of concern.
4. Easily detectable and distinguishable from other organisms in short period of time.
5. Growth should not be affected adversely by other food-associated microorganisms.

5.10.2. Faecal Coliforms

i. *E. coli* or Coliform Criteria

Presence of coliform or *E. coli* in foods is highly undesirable as it indicates faecal contamination, and handling and processing of food under unsanitary conditions.

However, it is usually impossible to eliminate them in all foods (fresh/frozen). Therefore, limits for coliform have been set for certain sensitive foods where coliforms are permitted in low numbers ranging from 1 – 100/g or 100 ml. These values are considered feasible and expected to ensure product safely. Although coliforms are widely used in shellfish sanitation programme, these often fail as good predictors of sanitary quality because of presence of pathogens in waters meeting coliform standards. However, coliform criteria have been widely used in most foods as good indicator of fecal contamination.

ii. *Faecal streptococci*

Fecal streptococci as Indicators

Faecal streptococci are a Gram positive bacteria associated with human intestine and thus encountered in fecal matter with their numbers variying from 105 to 108/g of feces. Species of streptococci such as *Enterococcus faecalis* and *E. faecium* are commonly associated with feces. These are fastidious in their nutritional requirements and grow over a wide pH range than other food borne bacteria. These being microaerophilic grow in reduced Eh condition. Generally do not multiply in water especially when organic matter load is low. Compared to *E. coli* these are less numerous in human faces. Ratio of 4 or higher between *E. coli* and streptococci is considered as indicative of fecal contamination. Enterococci die off at slower rate than coliforms in water. Live longer in water than coliforms thus outlive the pathogen in water. These are found to be better indicators of food sanitary quality in frozen foods as they are more resistant to freezing condition. Also exibit resistances to adverse environmental conditions, and have better survival in dried foods than coliforms. These show poor relation with the incidence of food borne intestinal pathogen. Though they possess qualities suitable to be considered as indicator organisms they have not gained importance, as coliforms.

5.10.3. Total Viable Counts as Indicators of Sanitary Quality

The natural counts of microorganisms vary depending on the environment from which the fish is caught. Besides, microorganisms are added during processing and handling of food. Number of microorganisms present in fish starts increasing once the fish dies unless steps are taken to halt their multiplication. As the quality deterorates with increase in microbial count, the total viable counts are used as indicators of product quality. Thus, to ensure quality, limits or standards are set for total viable load for fresh and processed foods. Further, Increase in microbial build up in processed foods during storage is attributed to likely changes in storage/ preservation conditions. Total viable counts are of greater value as indicators of existing condition of given product than as predictors of shelflife, because the portion of counts responsible for ultimate spoilage are difficult to ascertain.

Microbiological Standards and Criteria

Control on the microbiological quality of food is ensured by the regulatory agencies and food industry. This is done to ensure the safe of food to consumers, extend shelflife and to ensure product consistency from one batch to another in terms of safety and shelf-life.

Distinguishing the food of acceptable quality from food of non – acceptable quality is made possible with the application of microbiological criteria.

The Purpose of Microbiological Criteria

- The purposes of microbiological criteria for foods are to give assurance that;
- The foods will be acceptable from the public health stand point. That is, the food will not be responsible for the spread of infectious diseases or for food poisoning.
- The foods will be of satisfactory quality. That is, the food will consist of good original materials that have not deteriorated or become contaminated during processing, packaging, storage, handling *etc.*
- The foods will be acceptable from aesthetic view point. That is, it is free from extraneous material such as hair, scales, plastics *etc.*
- The foods will have keeping quality that should be expected of product.

Establishing microbiological criteria for a food product is based on factors such as total number of organisms, number of indicator organisms and number or absence of pathogens.

Types of Microbiological Criteria

Three types of microbiological criteria have been defined by the International Commission on Microbiological Specifications for Food (ICMSF). They are:

- Microbiological standard
- Microbiological specification
- Microbiological guideline
- Microbiological standard

Microbiological standard is a criteria specified in a law or regulation for a specific food. It is a legal requirement that foods must meet and is enforceable by the appropriate regulatory agency. Microbiological standard is that part of the law or administrative regulation designating maximum acceptable number of microorganisms or specific types of microorganisms as determined by prescribed methods in a food produced, packed or stored or imported into the area of jurisdiction of an enforcement agency.

Microbiological Specifications

Microbiological specification is the maximum acceptable number of microorganisms or of specific types of microorganisms as determined by prescribed methods in a foods being purchased by a firm or agency for its own use. Microbiological specification is the criteria applied in commerce. It is a contractual condition of acceptance that is applied by a purchaser. Failure to meet the condition by supplier will result in rejection of the batch or a lower price.

Microbiological Guideline

A microbiological guideline is that level of bacteria in a final product that requires identity and correction of causative factors in current and future production or handling after production.It is used to monitor microbiological acceptability of a product or process. It differs from standard and specification in that it is advisory in nature rather than mandatory.

Factors to be included in Suggesting a Microbiological Criteria (ICMSF)

- A statement of the food to which the criteria applies. As foods of different origins, composition and processing provide different microbial habitats; different foods pose different spoilage and public health problems.
- A statement of the microorganisms or toxin of concern. These include spoilage and health aspects. A realistic decision has to be made based on microbial ecology of food in question.
- Details of the analytical methods to be used to detect and quantify microorganisms/toxins. Methods elaborated by international agencies, and scientific methods that confirm the compliance with guidelines to be used.
- The number and size of the samples to be taken from a batch of food or from source of concern such as a point in a processing line.
- The microbiological limits appropriate to the product and the number of sample results which must conform with these limits for the product to be acceptable.

Chapter 6

Techniqes Involved in Study of Microorganisms

Introduction

The raw, semi-processed and processed foods as well as the ingredients used in the preparation of processed foods may contain several microorganisms. The microbial examination of foods is generally done to know the presence or absence of specific microorganisms, types of microorganisms, number of microorganisms and their products in foods. The information on the microorganisms associated with food will help in; Estimating shelf life (suitability for human consumption) by determining microbiological quality of a food or constituent of food.Identifying the cause of spoilage or presence of pathogens, when food is implicated in food borne illness.

To ensure the safety and quality of aquatic foods, microbiological assessment in necessary throughout the production chain *i.e.* from harvesting, processing and distribution. Microorganisms are the main cause of spoilage of fresh fish and shellfish. Food borne pathogenic bacteria may also enter through the production chain causing food-borne infections and intoxications. Identification and enumeration of microbial spoilage bacteria; and detection of food-borne pathogens are therefore very important for the fish processing industries and regulatory authorities. Conventional microbiological methods are still considered as standard methods, although they are laborious and time-consuming. Alternatively several rapid, indirect and molecular methods have been developed and they shall be used for microbiological examination of aquatic foods.

Conventional Microbiological Methods

The most usual technique for the enumeration of microorganisms is the colony count method on agar media. The bacterial population is determined by inoculating dilutions of microbial suspension on the surface of solid dry agar medium (spread plates) or mixing the test suspension with the liquefied agar medium (pour plates). The other culture technique known as Most Probable Number (MPN) is mostly used for the determination of Enteriobacteriaceae, coliform and *Escherichia coli*, when low numbers are suspected in the product. The conventional culture technique for the detection of pathogens includes mixing and homogenization of the product with general purpose pre-enrichment medium to help potentially stressed or injured cells to recover, subsequent selective enrichment into liquid medium to increase the population of the target microorganism, and finally plating onto selective agar media and confirmation of isolated colonies of the target microorganisms by phenotypic tests.

Enumeration of Total Viable Counts

Methods of Enumeration

Several methods are available for the enumeration of microorganisms from food. These can be broadly divided in to direct methods or indirect enumeration methods based on whether the microorganisms are counted directly or the products released by them is estimated.

I. Direct Enumeration Methods

a. Direct Counting Methods

There are several direct methods for the rapid detection of microorganisms from aquatic foods. They are direct epifluorescence filter technique, fluorescent in situ hybridization, flowcytometry, immunological and molecular methods.

- ☆ Direct microscopic count (DMC)
- ☆ Direct counting on membrane filters
- ☆ Flow cytometer

b. Culture Based Methods

- ☆ Plate count method
- ☆ Pour plate method
- ☆ Spread plate method
- ☆ Most probable number (MPN) technique

II. Indirect Enumeration Methods

a. Alternative Methods (Chemical and physical methods)

- ☆ Dye reduction test

- ☆ Electrical methods
- ☆ ATP determination
- ☆ Thermostable nuclease test
- ☆ Limulus Amoebocyte Lysate (LAL) test

b. Rapid Methods

- ☆ Immunological methods
- ☆ DNA/RNA methodology

Study of Microorganisms in Foods by Conventional Methods

Enumeration of microorganisms in foods is traditionally performed by directly counting all microorganisms present in foods or by allowing them to develop in to colonies and then counting.

6.1. Direct Enumeration Methods

6.1.1. Direct Counting Methods

Microorganisms present in the food can be counted by observing the food sample directly or retaining the microorganisms on a filter paper by filtering the sample and then observing under microscope.

A. Direct Microscopic Count (DMC)

DMC involves detecting the presence of microorganisms in food by microscopic observation. It is a simple method and easy to perform. DMC is performed by making a smear of food specimen/cultures on to microscopic slides, staining with appropriate dye and viewing and counting all cells using microscope under oil immersion objective. This method can be used only when microorganisms are present in large numbers (10^6/ml). It is commonly used in dairy industry for assessing microbial quality of raw milk and other dairy products.

Advantages

- ☆ Rapid and easy enumeration
- ☆ Can be employed to any foods (Ex: dried/frozen foods)
- ☆ Simple to perform
- ☆ Cell morphology can be assessed
- ☆ Efficiency can be increased by using florescent probes

Disadvantages

- ☆ Required dilution of sample

Direct Counting on Membrane Filters

1. Direct Epifluorescence Filter Technique (DEFT)

DEFT is a technique that relies on direct microscopy using the stain acridine orange which binds onto nucleic acid and fluoresces orange. After sample pre-treatment with enzymes and detergent, the microbial cells are concentrated on the polycarbonate membrane filter before staining with the dye. DEFT is a rapid method, but lacks sensitivity (10^4-10^5 cells/ml).

Membrane fillers with pore size smaller (0.45 um) than bacteria retain bacteria and the retained bacteria can be counted using microscope.

Procedure Involved

- Concentrating/collecting bacteria on polycarbonate filters by filtering known volume of homogenized sample.
- Staining and counting of retained bacteria.
- Placing the membrane on a nutrient agar media or absorbent pad saturated with culture media of choice, and incubating
- Following growth, colonies are counted

Advantages

- Well suited for samples containing low numbers of bacteria
- Facilities concertinaing bacteria by filtering large volume of sample
- Only small volume of food samples can be used for a single membrane
- Efficiency of membrane filter method can be increased by staining with florescent dyes (Ex. acridine orange) and observing under epiflorescence microscope (DEFT: Direct Epiflorescence Filter technique). Viable cells fluoresce green are counted. Non viable cells appear orange. Acridine orange is an metachromatic fluorochrome which binds to double stranded DNA of viable bacterial cells.
- Can be used to enumerate microorganisms from a variety of foods (fresh fish, meat, fish/meat products, water samples *etc.*).

2. Flow Cytometry

In flow cytometry, bacterial cells are forced to pass through a capillary in front of a focused light beam or laser. The instrument measures the forward and side scatter and fluorescence. Scattering is related to surface properties, while fluorescence is emitted by cellular components. Fluorescent dyes or other staining compounds or probes that can reveal the viability or metabolic state of the cell can be used. Flow cytometry can be used for automated cell counting and microbial cell detection. Despite the ability to measure individual cells, the handling of small volume makes flow cytometry unable to detect very low bacterial concentration. For microbial detection, an enrichment step is necessary. The total measurement time was about 30 min including preparation of the microbial cells for flow cytometry. Good correlation was obtained with the standard colony count methods in the range 10^5-10^9 cells/g.

6.1.2. Culture Based Methods

Culture methods involve examination of microorganisms in food by encouraging them to multiply in a liquid or solid media. On solid agar media bacteria develop as colonies and counting such viable colonies gives microbial load in foods. Enumerating microorganisms by culture based methods can be done by using plate count methods or MPN technique.

Culture Media

A wide variety of media with varied composition capable of supporting the growth are available for the cultivation of different microorganisms.

The composition of the media varies depending on;

1. Group/type of microorganism to be studied
2. Overall purpose of the study
3. Whether to grow wide range of microorganisms or specific types
4. Resusitation of damaged but viable cells
5. Type of diagnostic information required

Ex: General purpose media: Plate count agar Lactose broth: For *Escherichia coli* Seective media: Baird parker agarfor *Staphylococcus,* Bismuth sulphite agar for *Salmonella* TCBS for Vibrios.

A. Plate Count Method

This method is variously referred to as total plate count (TPC), standard plate count (SPC) or aerobic plate count (APC). It is most widely used conventional method for determining viable cells or colony forming units (CFU) in foods. SPC involves blending/homogenizing the sample, serially diluting in appropriate diluent, plating in or on suitable agar media, incubating at appropriate temperature for a given time, and counting visible colonies as CFU.

Principle Involved

SPC is based on the principle that each viable bacterial cells multiples and grows in to a visible colony. Thus, counting number of colonies gives an idea about bacterial cells present in a sample. Counts determined by taking average of replicate plates showing 30-300 colonies.

Factors affecting SPC

1. Sampling method employed.
2. Distribution of microorganisms in food.
3. Nature of food biota

4. Nutritional adequacy of plating media
5. Incubation temperature and time
6. Type of diluent used
7. Presence of other competing organisms *etc.*
8. Plating on selective media for specific organisms is limited by degree of inhibition and effectiveness of selective/differential agents employed.

SPC can be performed by pour plate method or spread plate method (surface plating method)

a) Pour Plate Method

Appropriate dilution of the sample (1 ml) is mixed with agar medium, allowed to set, incubated at appropriate temperature and colonies developed are counted. Here colonies develop both on surface and subsurface of agar plate. Proper mixing of sample with agar medium is necessary so as to get isolated colonies which can be done by 2 ways.

- ☆ One ml of appropriate sample dilution is added to petri-plate and about 15 ml of agar medium is added and mixed by rotating the plate in clockwise and anticlockwise direction.
- ☆ One ml of appropriate sample dilution is added to test tube containing about 15 ml of molten agar medium, mixed by rolling the tube between the palm and poured to petri-plates, allowed to set and incubated.

b) Spread Plate Method

Diluted sample (0.1 ml) is spread on the surface of pre-poured, hardened agar plates using glass rod, incubated at appropriate temperature and colony developing on surface counted.

Advantages

- ☆ Suitable for heat sensitive psychrotrophs in food as they do not come in contact with molten agar.
- ☆ Enables providing colony features useful in presumptive identification especially on selective media.
- ☆ Favours strict aerobes on surface, but microaerophils grow slowly.

Disadvantage

- ☆ Problem of spreaders and colony crowding makes the enumeration difficult.

B. Most Probable Number (MPN) Technique

MPN is suited for enumerating the presence of low numbers of microorganisms in foods. This involves inoculating replicate tubes of appropriate liquid media

(3 or 5 tube) with three different sample sizes/dilutions of the material to be studied and incubating at appropriate temperature. Then the absence or presence of growth is observed and MPN table consulted to get probable number of organisms in the sample. MPN numbers are generally higher than SPC.

Advantages

1. Relatively a simple method and assay to perform
2. Results are comparable from one laboratory to another
3. Specific group of organism determined by use of specific media.
4. Suitable for detecting organisms present in low numbers
5. Method of choice for coliform detection

Disadvantages

1. Requires use of large number of glassware and large volume of sample
2. Can not observe colony morphology
3. Lack of precision

Study of Microorganisms by Alternate and Rapid Methods

The types and number of microorganisms present in food sample can be determined either by measuring the metabolites released by the microorganisms and constituents of microbial cells employing physical/chemical methods, or by using rapid methods such as enzyme linked immunosorbant assay (ELISA) or polymerase chain reaction (PCR).

6.2. Indirect Enumeration Methods

6.2.1. Alternative Methods (Physical and chemical methods)

A. Dye Reduction Test (DRT)

Dye reduction test is one of the rapid methods for detecting the bacterial quality of foods. It is widely used to assessing the bacteriological quality of milk for a long time. The dye reduction time (DRT) has an inverse relationship with the bacterial load. Resazurin, methylene blue, and triphenyl tetrazolium chloride are the most commonly used dyes. Resazurin reduction time is 240 min, when the bacterial count is approximately 10^4 bacteria/ml and is 90-120 min, when bacterial count ranged from 1.5×10^6 to 7.70×10^6/ml. Methylene blue reduction test has been applied for assessing the bacterial load in shrimps and oysters. Tetrazolium reduction time is 360-390 min, when bacterial counts ranged between 1.50×10^6/ml. Resazurin test has been applied for microbiological control of deep frozen shrimps.

DRT is mainly used in dairy industry for assessing overall microbial quality of raw milk. The number of viable organisms in a sample is determined by their ability to reduce the redox dyes. The redox dyes take up electrons from active biological systems and are coloured when oxidized and coluorless when reduced.

Commonly Used Redox Dyes

Methylene blue: appears blue when oxidized, and colorless when reduced. Resazurin: appears blue when oxidized, and pink or white when reduced. Triphenyl tetrazolium salt: appers different from above two dyes in reduced/oxidized stste. It is colorless when oxidized, and coloured (red or maroon) when reduced due to formation of formzan.

Procedure

Food supernatant is added to standard dye solution, incubated (Ex. 10 min in resazurin dye) and color observed. Extent of reduction is related to bacterial load. Time for reduction to occur is inversely proportional to number of organisms present in the sample.

Advantages

- Simple, rapid and inexpensive method.
- Suitable for assessing quality of raw material at farm or dairy.

Disadvantages

- Not all organisms are able to reduce dye equally.
- Not suitable for food that contain reductive enzymes.

B. Electrical Methods

a. Impedance Method

Impendence is the total resistance to the flow of an alternating current through a given conducting material. Production of metabolite products changes the electric properties due to the production of charged molecules (*e.g.* lactic acid) from the catabolism of molecules that do not have an electrical charge (*e.g.* sugars). The time required to detect (T_d) a change of these properties, is inversely proportional to the logarithm of the initial microbial population. Change occurs until the microbial population levels reach 106cfu/ml. It is obvious that determination of the microbial population in the sample requires a calibration curve of T_d against plate counts.

b. Capacitance Method

Capacitance is a measure of the capacity of storing electricity charge by the substance. Capacitance is used to quantify *P. fluorescens* from rainbow trout, atlantic salmon, red snapper and tilapia fillets using Brain Heart Infusion (BHI) as a growth medium. Capacitance is used to quantify the total microbial population and evaluate microbiological quality of raw shrimps. The application of capacitance microbiology is attempted to detect and enumerate *V. cholerae* and *V. vulnificus*.

This method is a physical method and is one of the most widely used alternative methods for microbiological analysis of foods.

Principle growth of microorganisms in a liquid medium changes the chemical composition of the medium, which leads to changes in its electrical properties. Measuring changes in electrical properties forms the basis of determining the microbial load in a sample. This is done by measuring the electrical impendence. When microorganisms grow in a culture media, metabolite substances of low conductivity in to the products of higher conductivity, thereby decrease the impendence of the media. In broth culture, measurement of independence over time gives reproducible results for species and strains of microorganisms. It is capable of detecting organisms in the range of 10 to 100 cells. Generally, cell populations of 105-106/ml are detectable in 3-5 hours, and 104-105/ml in 5-7 hours. The times noted are required for the organisms to attain the threshold of 106-107 cells/ml.

Application

1. Useful in assessing the quality of vegetables: 90-95 per cent agreement has been found between impedence measurement and TPC, requires 5 hours to analyse and suitable for ground meat and other foods.
2. Microbiological quality of pasteurized milk was assessed by using impedence detection time of 7 hour or less, which is equivqlent to TPC of 10^4/ml or more bacteria.

c. Conductance Method

Conductance is the ability for electricity to flow through a substance. A sensitive and selective conductance method is developed for quantitative determination of *P. phosphoreumin* fresh fish and less than 50 cells/g of fish could be detected in less than 45 h. Metronidazole, carbenicilline, cetrimide, cycloheximide, diamide (MCCCD) as the conductance medium is used to quantitatively determine the *Psuedomonas* spp. population in European sea bass. *E. coli* in bivalve shellfish is quantified successfully in 96 per cent of cases, and only 0.7 per cent of *E. coli* cultures failed to give a conductance signal within 20 h.

C. ATP Measurement

Bacterial ATPase Activity

One of the first attempts to employ indirect technique in seafood analysis is the determination of adenosine triphosphate (ATP), to estimate the total number of viable cells in fish. Bacterial ATP is extracted and measured by the luciferin/luciferase bioluminescent reaction. The intensity of light emitted is directly proportional to the quantity of ATP and therefore the number of viable cells. This rapid method can detect populations as low as 102-103 bacterial cells in less than 1 h. However, the results are not consistent due to the different amount of ATP per cell for different microorganisms or different physiological status. Additionaly, background ATP from food material, biological fluids, *etc.*, interferes with microbial ATP. This method is now widely used for the monitoring the cleaning and disinfection of food processing surfaces.

Adenosine triphosphate (ATP) is the primary source of energy in all living cells and universal agent for the transfer of free energy from catabolic process to anabolic process. ATP generally disappears within 2h after cell death. The amount of ATP per cell is generally constant (10-18 – 10-17 mole per bacterial cell which is equivalent to 4 x 104 M ATP/105 cfu of bacteria). ATP in exponentially growing bacterial cells is 2-6 nanomole ATP/mg dry weight (about 0.40 per cent of dry weight of bacteria). Thus the measurement of cellular ATP can be equated to individual groups of microorganisms. A linear relation is observed between microbial ATP and bacterial numbers.

Procedure

ATP measurement is done by using firefly luciferin- luciferase system. In the presence of ATP, luciferase emits light which is measured by luminometer. Amount of light produced is directly proportional to the amount ATP added. Luciferase produces one photon of light on hydrolysis of 1 ATP molecule. ATP facilitates the formation of enzyme – substrate complex which is oxidized by molecular oxygen.

Application

ATP assay is proved to be suitable in foods for assessing microbial quality, and involves complex sample preparation procedures. One of the problem in ATP assay in foods is contribution from non-microbial ATP mainly from foods which can be removed by sample processing procedures. In meat non-microbial ATP removal involves centrifugation, use of cation exchange resin and filtration.

Though this is a rapid and sensitive method not used for routine monitoring of microbial contamination in foods. However, suitable for monitoring hygiene in food processing plants. On the spot monitoring of food handling surfaces/equipments can be done by swabbing a designated area and reading the relative light units (RLU) using lumonometer. The amount of ATP measured is of both microbial and non-microbial origin and the presence of both indicates poor hygiene. Thus it is valuable for monitoring purpose but not for indicating numbers of microorganisms.

D. Thermostable Nuclease Test

It is a chemical method used to detect presence of *Staphylococcus aureus*, a food poisoning organism in foods by detecting thermostable nuclease. *S. aureus* involve in food poisoning by producing enterotoxin which is a neurotoxin.

The enterotoxin producing *S. aureus* produce coagulate and nuclease enzyme, and a high correlation is observed between these two enzymes. Nuclease of *S. aureus* is more heat stable than nuclease of other Staphylococcus sp and other bacteria. Coagulase is not heat stable and hence it is not used. Increase in cell numbers increases the extractable thermonuclease of *staphylococcal* origin. Presence of 0.34 units of nuclease corresponds to 9.5 x 10-3μg of enterotoxin by *S. aureus*. Thermostable nuclease assay was found to be as good as coagulase assay for toxigenic strains. All the foods that are contaminated with enterotoxin were found to contain thermostable nuclease at *S. aureus* level of 106/g of food. Nuclease is detectable in sample at *S. aureus* numbers of 105-106/ml. And enterotoxins is detectable at cell number >106/ml.

Advantages

1. Thermostable nuclease is heat stable hence persists in food even after the bacteria are destroyed by heat/chemicals *etc.*
2. Thermostable nuclease is detectable faster (within 3 hrs) than enterotoxin.
3. Nucleases are produced by enterotoxignic stains before the appearance of enterotoxin.
4. Nuclease estimation does not require concentration of cultures in food, but enterotoxin detection requires concentration of samples.
5. Nucleases are heat stable, like enterotoxins.

E. Limulus Amoebocyte Lysate (LAL) Test

Limulus Amoebocyte Lysate (LAL) is a highly sensitive and rapid test used to measure the endotoxin producing gram negative bacteria in foods. Mostly gram negative bacteria (GNB) are responsible for the spoilage of raw fish. The gram negative bacteria except *Shigella* dysenteriae and *Vibrio cholera* are characterized by their production of endotoxins, which consist of lipo polysaccharide (LPS) layer on the cell envelop. LAL is an aqueous extract of the blood cells called amoebocytes from the horseshoe crab, *Limulus polyphemus*. LAL employs a lysate protein obtained from the blood, which is most sensitive to endotoxins. The test is performed by the addition of food suspensions (0.1 ml) to the lysate protein followed by incubation at 37°C for 1 h. The presence of endotoxins causes gel formation of the lysate material. Since both viable and non-viable gram – negative bacteria are detected by the LAL test, simultaneous plating is necessary in order to determine the number of viable organisms. Endotoxin titres increases in proportion to the viable counts of gram- negative bacteria.

Limulus lysate test is a chemical method used to detect the presence of endotoxin in foods. Endotoxin is produced by pathogenic Gram negative bacteria which consist of lipopolysaccharide (LPS) layer and lipid A in their cell wall. LPS is pyrogenic and responsible for ssymptoms associated with infection by Gram negative bacteria. Limulus amaebocyte lysate (LAL) test uses lysate protein obtained from the blood (haemolymph) cells (amaebocytes) of horse shoe crab, Limulus polyphemous. The lysate protein is most sensitive substance known for endotoxin. LAL test is performed by adding aliquots of food suspensions to small quantities of lysate preparation and incubated at 37°C for 1 hour. Presence of endotoxin causes gel formation of lysate material. LAL reagents can detect endotoxin as low as 1 pg of LPS. Incorporation of chromogenic substrate (p-nitroaniline), to endotoxin activated enzyme cleaves the substrate and releases free p- nitroaniline that can be read at 405 nm. The amount of chromogenic compound released is proportional to the quantity of endotoxin in the sample. By knowing the amount of endotoxin per cell of Gram negaticve bacteria (which is fairly constant), the total bacterial load can be determined from the quantity of endotoxin measured.

Application

LAL test is a good and rapid indicator of total number of Gram negative bacteria in refrigerated foods (fish/meat) which are spoiled mainly by Gram negative bacteria

Advantages

1. LAL test detects both viable and non-viable Gram negative bacteria in food sample.
2. Found suitable for assessing microbial quality of milk/milk products and raw fish.
3. Gives quick result.

Food of high LAL value need further tests by other methods. Foods with low LAL titre can be categorized as low risk relative to Gram negative bacteria.

6.2.2. Enumeration of Microorganisms by Rapid Methods

The term "rapid methods" refers to improved, alternative methods that can enable early, accurate and faster detection, isolation, characterization and/or enumeration of microorganisms. There are direct and indirect rapid methods available for the detection of microorganisms from aquatic foods.

The conventional cultured based methods often fail to detect the presence of microorganisms in foods and also take long time for laboratory analysis. This may be due to their low cell numbers and loss of viability due to damage to cells. This can be overcome by applying culture independ methods involving the detection of bacterial cell components and metabolites by immunological and molecular based approaches. The commonly employed rapid tests in food microbiology are;

A. Enzyme linked immunosorbant assay
B. Polymerase chain reaction

A. Enzyme Linked Immunosorbent Assay (ELISA)

Methods based on highly specific antigen antibody binding have become widespread during recent decades for detection and/or identification of food-borne pathogens. Specialized monoclonal or polyclonal antibodies bind on specific sites on cell surfaces allowing visualization of the reaction through precipitation or other means, such as colour or fluorescence development.

The most widely used immunological technique is the ELISA. In general, the target antigen of the sample binds to the antibodies that are fixed on the surface of microtitre plate wells. After washing to remove any unbound antigen, enzyme labeled antibodies are added and target the antigens in the wells. The visualization is possible, after washing to remove any unbound antibodies, followed by the addition of a substrate that the antibody linked enzymes converts to a visual signal. The method is rapid, but the detection limit is about 10^5-10^6 cells/ml, which corresponds to 10^6-10^7 cells/g in food. This is a reason that immune-based techniques have not been used for determination of spoilage microorganisms despite the development of antibodies for S. putrefaciens. However, it can be used for pathogen detection in seafood. An enrichment step is required, which increases analysis time, but detection of food-borne pathogens can be substantially decreased down to 1 or 2 days, compared to 5 days or more required with conventional culture techniques. Many ELISA tests are automated in order to reduce the hands-on time and increase reproducibility.

ELISA or enzyme immunoassay (EIA) is used in food microbiology to identify the specific pathogens or toxins released by them. Detection of specific microorganism among the mixture of organisms is made easier by employing ELISA.

This method based on the antigen (cell or toxin) – antibody reaction is highly specific and helps to detect when antigens are present in very low levels. Thus, ELISA involves specific reaction between the antigen, antibody and an enzyme. This reaction complex produces a colour in the presence of chromogenic substrate which is proportional to the amount of antigen present. The enzymes such as horse radish peroxidase and alkaline phosphotase are commonly used which release a dye (chromogen) when exposed to their substrate.

Procedure

- ☆ The antigen to be detected is taken in a test tube or microtitre plate and incubated with the antiserum.
- ☆ The excess antiserum is washed and the enzyme labelled with anti-immunoglobulin is added.
- ☆ After washing, the enzyme remaining in the tube or microtitre well is assayed to determine the amount of specific antibodies in the initial serum.
- ☆ The amount of peroxidase enzyme is measured by adding enzyme specific substrate and the colour developed is measured colorimetrically.

B. DNA/RNA Methodology

Polymerase Chain Reaction (PCR)

Polymerase chain reaction (PCR) is a molecular biology technique employed in food microbiology to identify the presence of specific microorganism of interest especially when present in very low numbers by targeting the specific gene sequence. As the foods may contain pathogenic microorganisms which may be present in low numbers or injured by conditions of food processing, the recovery of such organisms by routine plating techniques is not possible. In such situations PCR based methods become very useful in identifying the target organism. The

PCR for the detection of pathogens generally targets virulence associated genes such as ctx gene encoding cholera toxin for *Vibrio cholerae* 01 and 0139; tdh gene coding thermostable direct hemolysin for *V. parahaemolyticus*; hns and invE gene for *Salmonella*; stx1, stx2 and eae gene for enterohaemorrhagic *E. coli*.

Principle of PCR

The method is based on the amplification of target DNA sequences in the genomic DNA in the presence of specific primers, oligonucleotides, buffers and polymerase enzyme. Under the conditions of repeated heating and cooling cycles millions of copies of target DNA will be synthesized with in a short period of time. The amplified product is detected after performing electrophoresis on agarose gel and visualized under UV light after staining with ethydium bromide.

Advantages

- PCR is highly specific as it can amplify a target DNA fragment of pathogen of interest against DNA of other organisms.
- PCR is highly sensitive as it is possible to amplify target sequence from samples having very few microbial cells of interest.
- PCR is a rapid test since the results can be obtained with in a few hours.
- It facilitataes use of food samples directly or after enrichment. Sample lysate or enrichment lysate can be use for extracting DNA for further PCR amplification.

Molecular Methods

Molecular methods that are gaining popularity in food safety assurance are polymerase chain reaction (PCR) and DNA probe hybridization methods. The former is based on nucleic acid amplification and has very high sensitivity, but requires special laboratory facilities and can also detect dead bacteria in processed foods. On the other hand, DNA probe hybridization, performed based on colony hybridization, does not require expensive equipment, but detects only live bacteria and gives quantitative data. There has been a tremendous development in the past 15 years with the introduction of molecular methods for the detection and identification of microbial pathogens, including bacteria, virus and protozoa. There are many DNA-based assay formats, but only probes and nucleic acid amplification techniques are commercially applied for pathogen detection.

i. Conventional Polymerase Chain Reaction (PCR)

Polymerase chain reaction (PCR) is a method used for the *in vitro* enzymatic synthesis of specific DNA sequences by Taq and other thermoresistant DNA polymerases. PCR uses oligonucleotide primers that are usually 20–30 nucleotides in length and whose sequence is homologous to the ends of the genomic DNA region to be amplified. The method is performed in repeated cycles, so that the products of one cycle serve as the DNA template for the next cycle, doubling the number of target DNA copies in each cycle. The rapid increase in the number of copies of the target sequence that can be achieved with PCR-based methods makes them ideal candidates for the development of faster microbiological detection systems. Many PCR tests have been validated and commercialized to make PCR a standard tool used by food microbiology laboratories to detect pathogens in foods. Conventional PCR relies on amplification of the target gene(s) in a thermocycler, separation of PCR products by gel electrophoresis, followed by visualization and analysis of the resulting electrophoretic patterns, a process that can take a number of hours. The specificity can be subsequently confirmed by sequencing the amplifyed fragment. PCR can be superior to culture for detecting the main pathogens in food samples. PCR is widely used for detection seafood associated pathogens.

ii. Nested PCR

Nested PCR is increases the specificity of DNA amplification, by reducing background due to non-specific amplification of DNA. Two sets of primers are used in two successive PCRs. In the first reaction, one pair of primers is used to generate DNA products, which besides the intended target, may still consist of non-specifically amplified DNA fragments. The product(s) are then used in a second PCR with a set of primers whose binding sites are completely or partially different from and located 3' of each of the primers used in the first reaction. Nested PCR is often more successful in specifically amplifying long DNA fragments than conventional PCR, but it requires more detailed knowledge of the target sequences.

iii. Multiplex-PCR

Multiplex-PCR consists of multiple primer sets within a single PCR mixture to produce amplicons of varying sizes that are specific to different DNA sequences. By targeting multiple genes at once, additional information may be gained from a single test-run that otherwise would require several times the reagents and more time to perform. Annealing temperatures for each of the primer sets must be optimized to work correctly within a single reaction, and amplicon sizes. That is, their base pair length should be different enough to form distinct bands when visualized by gel electrophoresis.

iv. Reverse Transcription PCR (RT-PCR)

Reverse transcription polymerase chain reaction (RT-PCR) is one of many variants of polymerase chain reaction (PCR). This technique is commonly used in molecular biology to detect RNA expression. It amplifying DNA from RNA. Reverse transcriptase reverse transcribes RNA into cDNA, which is then amplified by PCR. RT-PCR is widely used in expression profiling, to determine the expression of a gene or to identify the sequence of an RNA transcript, including transcription start and termination sites. If the genomic DNA sequence of a gene is known, RT-PCR can be used to map the location of exons and introns in the gene. The 5' end of a gene (corresponding to the transcription start site) is typically identified by RACE-PCR (Rapid Amplification of cDNA Ends).

v. Real-time PCR

Real time PCR (also called quantitative PCR or qPCR) is now becoming popular for the detection of food -borne pathogens due to the possibility of quantitating them and eliminating the need for gel -based detection of PCR products. There are two methods of quantifying the PCR products: Use of fluorescent dyes that intermediate with double stranded DNA and use of modified oligonucleotide probes that fluoresce when hybridized with complementary DNA. Dyes such as SYBER Green that bind dsDNA would bind to all dsDNA, including non-specific products or primer dimers, but are less expensive and can be used for any target to be amplified. The fluorescent reporter probes, on the other hand, need to be specifically synthesized for each reaction.

The TaqMan assay in which a single stranded oligonucleotide probe complimentary to a segment of 20 to 60 nucleotides, within the DNA template and located between the two primers is used. In this assay, a fluorescent reporter (*e.g.* 6-carboxy fluorescin) and quencher (*e.g.*tetramethyl rhodamine) are covalently attached to the 5 and 3 ends of the probe, respectively. The single stranded probe does not show fluorescence due to close proximity of fluorochrome and quencher. During PCR, the 5 to 3 exonuclease activity of Taq polymerase degrades the portion of the probe that has annealed to the template, releasing the fluoro chrome from proximity to the quencher. This fluorescence is directly proportional to the fluoro phore released and amount of DNA template present in the PCR product. With both types of assays, the exponential increase in fluorescence is used to determine the cycle threshold(Ct), which is the number of PCR cycles at which significant exponential increase in fluorescence is detected. Using a standard curve for Ct values at different DNA concentrations, quantitation of target DNA in any sample can be made.

Vi. DNA Microarray

DNA microarray technique is a sensitive, rapid test for microbial quality assurance of fish and fishery products. The microarray method of analyzing bacterial pathogens has potential application in fish and fishery products safety as a rapid detection method of multiple bacterial pathogens. The term DNA microarray derives from the array of orderly arranged probes that are spotted and immobilized (chemically bonded) on a solid matrix (*viz.* glass slide) with the help of some immobilizing materials. Hundreds to thousands of probes are being spotted onto the solid matrix. DNA samples to be analysed are labeled with some fluorescent dyes and allowed to hybridize onto the already known probes on the solid matrix, relying on the principle of base pairing. The fluorescent-labeled DNA samples that hybridise with the known probes are thereby identified by the scanning signal of intensity of fluorescence. This technique can be used for the recognition of oligonucleotides, complementary DNA (c DNA), protein, peptide, antibody, *etc.* One of the applications of DNA microarray in pathogen detection is microbial genotyping. Microarray genotyping is used to determine whether a specific sequence is present in a genome or not. However, in food –borne pathogen analysis, microarray genotyping is more commonly used to identify limited number specific sequences in a food sample for the purpose of bacterial pathogen identification or differentiation between the bacterial strains.

vii. DNA Probe Hybridisation Methods

DNA probe hybridization is based on the principle that, the two strands of double stranded DNA are separated by heating or chemical treatment, which these reassociate and hybridise, provided there is complementarity of bases (A-T; G-C) between them. Based on this principle, it is possible to make probes specific to different microorganisms. Probes are short nucleotides that have sequences complementary to the target sequences. Probes are labeled either with a radioactive molecule, enzymes, ligands (*e.g.* biotin) or antigenic substrates (*e.g.* digoxygenin) to detect hybridization. Figure illustrates a typical probe.

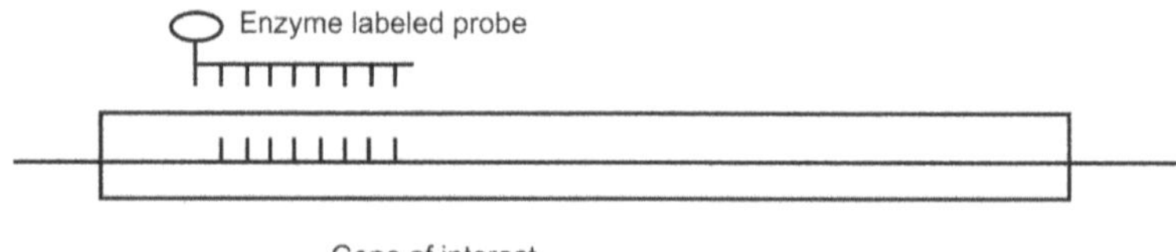

Illustration of the Principle of DNA Hybridization Technique.

The general protocol for detection of pathogenic bacteria by colony hybridization includes homogenization of fish sample in a buffer (1:1) and plating on a non-selective agar. The colonies appearing after incubation at 35°C for 18 h. are transferred to a suitable filter and lysed by immersion in a lysing solution. The DNA released are denatured and fixed to the filter the filter is then incubated in prehybridizing solution and hybridisates is performed at appropriate temperature after the addition of the probe. If radioactive probe is used, detection is by autoradiography. There are a number of non radioactive labels now available. The most convenient are enzyme label which can be detected using the chromogenic substrate.

To detect pathogens that are present in small numbers in the fish, is enrichment step is essential before plating. This analysis does not require any equipment except a hybridization incubator and can be conveniently adopted in quality control laboratories. DNA probe-based methods are required to detect pathogenic strains of *Vibrio parahaemolyticus*, *Salmonella* and *Listeria monocytogenes etc.*

Genes that are chosen as targets are those specific for each bacterium. In the case of pathogenic organisms, these are genes that encode virulence factors *i.e.* factors that make the organism pathogenic. Some examples of these genes are given in Table Pathogenic *V. cholera* produce a toxin, cholera toxin encoded by ctx gene. The virulence factors in *V. parahaemolyticus* are encoded by tdh and trh genes. The virulence factors in *L. monocytogenes* are encoded by iap and hly genes. In *Salmonella*, there are virulence associated genes such as inv and hns. Enterohemorrhagic *E. coli* have virulence genes such as stx, eae. Using such specific probes, it is possible to specifically detect the pathogenic strains of bacteria.

Examples of Pathogens and Target Genes

Pathogen	*Target genes for probe hybridization*
Vibrio cholerae	Ctx
V. parahaemolyticus	tdh, trh
V. vulnificus	Vvh
Salmonella	inv, hns
Enterohemorrhagic *E. coli*	stx, eae
Enterotoxigenic *E. coli*	St, Ct
L. monocytogenes	iap, hly

viii. Fluorescent In situ Hybridization (FISH)

FISH is a molecular technique, as well as a microscopy based method. In this method, oligonucleotide targeted probes labeled with a fluorescent dye hybridize the target RNA of the microbial cells that are chemically fixed on a glass slide. After washing to remove any unbound probes, the stained cells are detected using epifluorescence microscopy. The detection limit is about 104 cells/ml. The practical application of FISH technology is more difficult in food compared to water samples, as food particles interfere with detection of the bacterial cells by epifluorescence microscopy. For the detection of low bacterial numbers, enrichment is necessary. FISH filter cultivation (FISHFC) method, which includes a brief cultivation step on a membrane filter before in situ hybridization is rapid (7h) for the enumeration of Enterobacteriaceae in food samples at population levels as low as 10^2 cells/g. With the FISHFC technique, detected and enumerated Listeria spp. in smoked salmon in less than 14 h.

6.3. Encapsulation – Endospores, Formation of Cell Aggregates

Endospores and Formation of Cell Aggregates

Endospores are metabolically dormant stages observed in certain Gram positive bacteria as a survival strategy to overcome unfavourable environmental conditions. Spore forming bacteria are of significance in food industry because of their ablity to cause food spoilage and produce toxins which cause illness in humans. Among the bacteria, *Clostridium botulinum, C. perfringens* and *Bacillus cereus* are toxigenic while many species of spore formers cause spoilage of food. Endospores are produced inside bacterial cell by members of Gram positive *Bacillus, Clostridium, Desulfotomaculum, Sporolactobacillus and Sporosarcina*. Spore formation also called as sporulation or sporogenesis is part of the natural life cycle of bacteria. Spores differ from metabolically active and growing vegetative cells by their inert resting condition. Endospores vary in size, shape and position in the vegetative cells in different bacteria and are often useful in the identification of some species.

6.3.1. Endosopre Formation

Endospore formation is initiated by the vegetative cell under the conditions of nutrient depletion, especially the carbon and nitrogen source. The vegetative cell prepares for sporulation by transforming in to a committed sporulating cell called sporangium. The sporangium actively involves in synthesizing compounds required for spore formation. Most spore formers develop mature spore of complex structure within 6-8 hours. Sporulation usually appears in the late logarthemic phase of growth possibly because of nutrient depletion and accumulation of toxic metabolites.

Germination of Spores

The spores remain in dormant state for a varying period of time, even for several years. When conditions become favourable the spores break dormancy and enter in to a process called spore germination. Presence of water and certain specific chemicals such as aminoacids or inorganic salts and environmental stimulus initiate germination process by activating dormant hydrolytic enzymes from spore membranes. These enzymes digest the spore cortex and expose the core to water. The rehydrated core utilizes nutrients and grows out of spore coat fully reverting back to vegetative cell. Generally spores germinate in to vegetative cells within a short period of about 90 minutes.

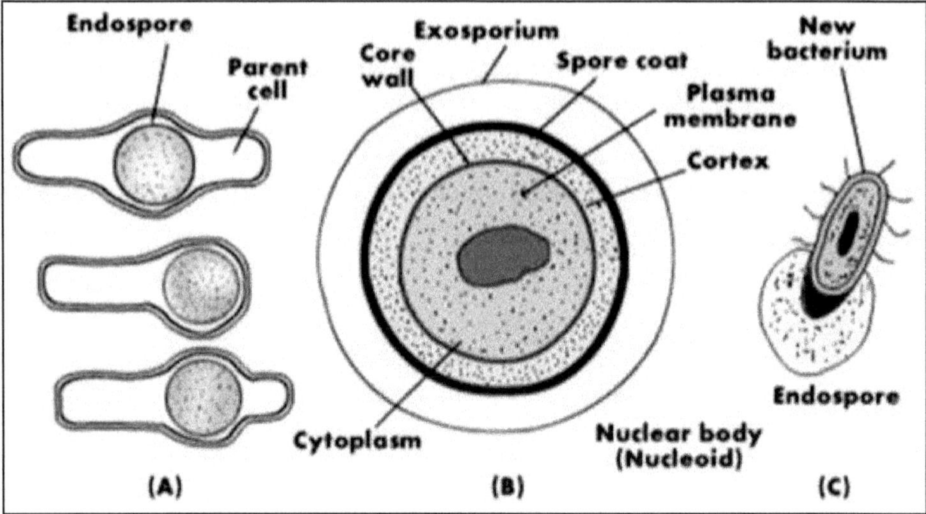

Endospore Formation. (A) Endospores accoring to their position in parent cells. (B) An endospore in cross-section. (C) Germination of endospore.

Resistance of Spores

Bacterial endospores are highly hardy structures capable of withstanding extreme heat, drying, freezing, radiation and chemicals that would readily kill vegetative cells. Ability to survive such harsh environmental conditions is attributed to several factors.

- The high heat resistance of spores is due to high content of calcium and dipicolinic acid which removes water and makes it dehydrated. The absence of free water offers heat resistance and thus protective effect on proteins and nucleic acids.
- The spore is metabolically inactive due to non availability of water which makes it resistant to further drying.
- Presence of thick and impervious cortex and spore wall offer resistance against radiation and chemical substances.

6.3.2. Cell Aggregates

Microorganisms occur either as single cells or as aggregates consisting of clumps or chains of cells. Microorganisms always grow attaching on to suspended organic and inorganic substances rich in nutrients. In food industry spoilage and pathogenic organisms grow on food residues and serve as continuous source of contamination to food. Also, the bacterial cell aggregates are more resistant to cleaning and disinfection in food industry. As the microbial cells occur in several layers on food residues and surfaces containing nutrients, the cells in the inner layers are not exposed to disinfection treatment and hence survive better. Formation of bacterial cell aggregates need to be prevented in any food handling and processing environment by following good sanitary measures.

References

Alasalvar, C., Shahidi, F., Miyashita, K., Wanasundara, U. 2011. Handbook of Seafood Quality, Safety and Health Applications. Blackwell Publishing Ltd. 275-284.

Duflos G., Dervin C., Malle P. and Bouquelet S. (1999). Use of Biogenic Amines to evaluate spoilage in Plaice (*Pleuronectes platessa*) and Whiting (Merlangus merlangus). *Journal of AOAC International*, 82 (6), 1357-1393.

Edema, M.O., A.M. Omemu and M.O. Bankole, 2005. Microbiological safety and quality of ready to eat foods in Nigeria. In: the book of abstract of the 29th Annual conference and general meeting (Abeokuta, 2005) on microbes as agents of sustainable development organized by Nigerian society for Microbiology (NSM), University of Agriculture.

Fernandes, R. and Wareing, P. 2009. HACCP in Fish and Seafood Product Manufacture. In: Fernandes, R. (Ed.) Microbiology Handbook Fish and Seafood. Leather Food International.

Gopakumar, K. 2006. Textbook of Fish Processing Technology. ICAR.

Greening, G.E, 2006. Human and animal viruses in food (including taxonomy of enteric viruses). In: Viruses in foods, (Ed., Goyal, S.M). Springer Science+Business Media, USA., pp. 5-42.

Iddya Karunasagar, Indrani Karunasagar and H. Sanath Kumar. 2002. Molecular methods for rapid and specific detectionof pathogens in seafood. *Aquaculture Asia*, Vol.VII No. 3, 34-36.

Jeyasekaran G., Jeya Shakila R. Shalini R. and Yesudhason P. 2015. National Training on Microbial Risk Assessment in Aquafoods, TNFU.

Jeyasekaran, G., Jaya Shakila, R., and Sukumar, D. (2006). Quality and safety of seafoods. TANUVAS.

Levin, R.E. (2010). Assessement of seafood spoilage and the microorganisms involved, in Handbook of seafood and seafood products analysis (eds L.L. Leo and F. Toldra), CRC press, Taylor and Francis Group, Boca Raton, FL, USA, pp. 515-535.

Mavromatis P. And Quantik P. 2002.Histamine toxicity and Scombroid fish poisoning: a Review.In. Alasalvar. C and Taylor.T Eds. Seafoods- Quality, Technology and Neutraceutical Applications 89-99.

Rhea Fernandes, 2009, Microbiology Handbook, Fish and seafood, Leatherhead Food International Ltd, Randalls Road, Leatherhead, Surrey KT22 7RY, UK, USFDA guidelines Chapter 7 Scombrotoxin (Histamine) Formation.

WTO, (World Trade Organisation), 2002. Committee on Sanitary and Phytosanitary Measures, G/SPS/N/MEX/189, Notification of emergency measures, World Trade Organization, New York, USA.

www.ingramcontent.com/pod-product-compliance
Ingram Content Group UK Ltd.
Pitfield, Milton Keynes, MK11 3LW, UK
UKHW021951270726
14060UKWH00002B/454